THE LONG UNWINDING ROAD

by

Jon Stamford

ACKNOWLEDGEMENTS

Lots of people really but mainly my family. As it should be.

PARKINSON'S DISEASE

Parkinson's disease, or simply Parkinson's as it is more commonly mentioned, is a progressive neurodegenerative disease attacking primarily the brain's dopamine systems.

It's more complicated than that obviously but the main things you need to remember are that it affects your ability to move and that it gets worse. And if that wasn't enough to trouble you, then the fact that it is incurable probably answers that point.

Before you get the impression, perhaps understandably, that this is all doom and gloom let me tell you that it is not.

Firstly there are drugs to treat Parkinson's. They're not perfect, and some carry significant side effects. But nothing in life comes without a price and these drugs work reasonably well in the early stages of the illness. In the later stages... well... not so well but then there are different drugs for that part of the illness.

One of the things you learn in Parkinson's is that there are ways round most problems. And part of the experience of Parkinson's is in finding those neat little workarounds or "hacks" as people are fond of calling them. Some are simple expediency – if your right hand is shaking, use your left. Others are more complicated learnt responses to more challenging issues.

But with a willingness to be open-minded and to learn from the experience of others, it is possible to live with this condition. So relax.

THE LONG UNWINDING ROAD

Before you get the wrong idea, may I say that this is not explicitly a book about Parkinson's. There is a whole bundle more stuff in it than that. Parkinson's is a theme certainly – I have Parkinson's and with the best will in the world it's hard to shake off its influence on my daily life. But the book goes much beyond there.

I have divided the book broadly into three years. This is mainly to give a context to some of the more ephemeral pieces that would otherwise not make sense such as the piece on Brexit in 2016.

That probably gives you also a clear understanding that there will be some politics amongst the pieces. Don't be put off. I have balanced it with other less knuckle whitening subjects. There is also a distinct whiff of nostalgia in some of the pieces as I recall my Yorkshire roots. And as if that was not enough on its own, I've even treated you, dear readers, to some of my thoughts on Wagner's music. You lucky people.

JON STAMFORD

Dr Jonathan (Jon) Stamford is a semi-retired neuro-scientist with Parkinson's. For more than a decade he directed a research lab looking at the actions of drugs in Parkinson's before being diagnosed with the condition himself in 2006.

Dr Stamford was among the first group of ambassadors for the World Parkinson Congress in Montréal in 2010 and was, until recently, Scientific and Advocate Communication Coordinator for the Cure Parkinson's Trust and Director of Parkinson's Movement.

He is co-founder of the Parkinson's Inside-Out group and a research consultant for Spotlight YOPD.

INDEX

2016

A NEW VOCABULARY

I often think the way we talk about Parkinson's disease has a significant bearing on how we live with the condition. Let me give you an example. The metaphor most commonly used is that of a journey. We talk about where people are on that journey. And I for one have always been deeply uncomfortable with the journey as a metaphor. Firstly, it implies direction. And a fixed direction at that. For me it conjures up images of train tracks, linear and unbranching, heading towards the horizon and, over that horizon, an unattractive and unnamed destination.

Slowly, insidiously this imagery begins not to describe the condition but to direct it. The more we think of a journey, the more our experience of Parkinson's begins to fulfil that description, the more the Parkinson's takes on the character of the journey.

This stands at odds with most people's objectives. In terms of Parkinson's, we want to stay still. We don't want to be moving. We need to see stasis as an appropriate aspiration. We need to celebrate inertia rather than movement.

That brings me to another aspect. Progression. We live in a world where progress and progression in life are desirable. Nobody has ever said "let's turn back progress".

But in Parkinson's, progress is bad and progression is merely a descriptive expression of that badness. In Parkinson's, it implies movement and movement in the direction of that destination that none of us wish to talk about.

When we meet with our neurologists, and they tell us they are happy with our progress, they mean the opposite. They mean they are happy with our lack of progress, with our symptomatic inertia. If your Parkinson's has progressed, this is not something to be applauded.

We need a new vocabulary and we need new imagery to inform our perception of this condition. We need to celebrate inertia. We need to celebrate stillness, calm and immobility. We have to rid ourselves of this notion of the journey. There is no predetermined journey. Nobody says we have to board the train. Why not take a seat on the platform, drink coffee and read the newspaper.

THE MAGIC WAND

It started out more as a symbol than as a practical aid. I have been aware that my walking has deteriorated over the last six months. Indeed that and a substantial increase in tremor are the principal outward manifestations of a progression of my Parkinson's.

The tremor is a nuisance but it does at least mark you out as different. Unsteady gait on the other hand is so often dismissed as the product of too fond a liking for the fermented grain or grape. Eventually it becomes tiresome, having to either endure or challenge the comical prejudices of outsiders. It takes too long to explain to every tittering halfwit that you are not drunk but have Parkinson's disease. You need some kind of shorthand. And for me, the best way of avoiding critical or derisive comment is to carry a stick.

The stick legitimises the most Python-esque of silly walks. No matter how unsteady I am, lurching or half running, stumbling and staggering, the stick says "it's okay, the owner of this stick is allowed to walk like this". And that's how it started out – one particularly unsteady week and one too many people laughing at my tiptoe shuffling walk were enough. I resolved to carry a stick from then on.

A stick opens doors. Literally. People will hold the door for you where previously they would have happily let it close in front of you. People stand aside to let you get on the train first where previously they might have brushed you aside or tut-tutted at your selfish impediment to their progress.

The stick finds seats. Whereas formerly I would stand on commuter trains when no seats were available, now I find I am offered a seat. Instead of facing a phalanx of defiantly raised broadsheets, I am given a seat. Sometimes more than one! People practically fall over themselves to help. And all because of my stick.

My disability is neither greater nor lesser because of the stick. I do not go from able-bodied to cripple simply because I carry a stick. Except of course in people's perception. Where I have in essence a badge, the stick, I am someone to be treated more kindly. And the converse is of course also true. Where I have no stick, I have no right to a seat on a train, no right to expect people to open doors and no right to get in your way as you hurry onto your commuter trains home.

But what does this say about us and society? For me, it highlights not the helpfulness of people toward the disabled so much as society's indifference (intolerance even) of those whose disabilities are not immediately visible. In other words, many people with Parkinson's. Do we have to carry a stick to mark us out as different and worthy of your help?

And where once it was a symbol of my infirmity, now it is a part of me. I travel with it more often than not. What once was token is now another limb. And like my blue badge, it is another unwanted Rubicon crossed. But also a major help. And I've learnt that whatever the motivation, help is not something to rail against. Grumpy rejection of offers of assistance help no one.

Rather like Harry Potter I did not choose my stick – it chose me. I was clearing out cupboards at my father's house after his death and, as I opened one under the stairs, a stick fell into my hand. Made of oak and half a century old it had belonged to my parents and grandparents before them.

It's not a walking stick. Think of it as a magic wand.

STEP BACK

I'm not by nature a political animal. My attitude to politicians is broadly speaking the same as double glazing salesman, ambulance chasers and PPI cold callers. I have long held the view that the mere desire to be a politician should automatically disqualify you from office. So, on the whole, I have little truck with politicians, local, national or global.

But every once in awhile an issue arises that transcends politics – an issue that threatens to change everything in life. I'm talking of course of Britain's EU referendum next week. One day in which we, the humble electorate, will take probably the most important decision over this country's future for a century.

So I'm forced reluctantly out of my apolitical cave and forced to play my tiny part in making that decision. Should Britain remain part of the European Union or should we leave it? This is a simple enough question. It's black or white and whichever shade of grey we may individually be, we are compelled, for the purposes of this poll, to be either one or the other. I resent that of course. I resent being forced to take a clear side. I want to discuss it and I feel rather as though I'm having an argument with someone who bangs on that all they want is a yes or no answer. I want to give my reasons. I want my interlocutor to understand why I think a certain way.

So do we remain in the union or secede?

It's often said that if we fail to learn the lessons of history, were condemned to repeat them. Try this for size: On 20 May 1861, following a vote, North Carolina formally seceded from the United States on the grounds that it placed self-determination above the legislative primacy of the union. It was preceded by Tennessee, South Carolina, Georgia and other states. Despite seceding from the union, North Carolina initially rejected joining the Confederate

States of America, only doing so when its borders were threatened.

Four years later, in April 1865, after one of the bloodiest conflicts in history with somewhere in the region of a million soldiers dying, North Carolina was readmitted to the union following Joe Johnston's surrender to General Sherman. Self-determination had been short lived and high-priced.

I'm being alarmist, right? The UK seceding from the European Union is different surely?

No. It is more similar than it is different. North Carolina seceded because of some fanciful notion of self-determination. This is national vanity of the first order. No nation exists in isolation. Indeed isolation is the route to nonexistence as a nation.

When North Carolina was readmitted to the union, it did so without any bargaining power at all. It was a country broken by the previous four years. Although it left the union with its head held high, it returned with its tail between its legs. In four years time, when our economy is broken by self-imposed isolationism, what terms will we be able to negotiate over our surrender. Will the European Union want such an abject apology for a nation state back in their midst? Will they welcome us back with open arms like prodigal sons?

Let's look at what might have happened in Europe in those four years.

Europe is by no means without its own inherent instabilities. It is well known that several other countries in Europe, including Denmark and Germany, are also flirting with the idea of referenda. The UK referendum is a litmus test and a decision to secede will almost certainly be followed by demands in other countries for a vote. Although we, as a nation, like to see ourselves at odds with mainland Europe, we are a respected nation within the European Union. Our opinion counts and, in many respects, we are one of the major reasons the union holds together. Some of the most important and valuable legislation in Europe such as the European Declaration on Human Rights was initiated by the British. To leave the union is to initiate a domino effect, with Denmark, Germany and probably other countries voting on whether to leave the union.

Some of the more extreme isolationists take the view that this is of no consequence. Britain will be an island nation, now immune to the vagaries of European politics.

This is risible. We were an island state in 1914. And who would have believed then that a single gunshot in the Balkans could draw Britain into the carnage of World War I and war against Germany. We were an island state in 1939. Chamberlain even secured an assurance that we would not find ourselves in conflict. Yet once again, instability in Europe forced us into war.

Britain leaving Europe will automatically destabilise the financial markets in the UK and Europe. Destabilise the financial markets and you destabilise governments. Couple that to increasing unrest with the notion of union and you have a powder keg. When you also factor in the significant lurch to the right seen in recent elections throughout Europe, the hairs on the back of your neck begin to rise.

Perhaps you think this is impossible. Perhaps you believe that a Britain outside Europe will control its borders better? Perhaps you feel we will be able to police our 2500 miles of coastline for the boatloads of refugees? Perhaps you didn't think about that – that war leads to refugees. Often in biblical numbers.

We have had no war within the European Union since its inception. If the union has achieved nothing else, that alone should be a source of significant pride.

When we go to the polls next week and the country takes the most important decision in the last hundred years. We stand on the edge of a cliff looking down. For goodness sake, step back.

JULY ON THE SOMME

On Friday 1st July 2016, we will celebrate (if celebrate is the right word) the centennial of the Battle of the Somme. The battle which raged, on and off, until November 1916, resulted in a million men either wounded or killed. Unimaginable numbers and a scale of slaughter which depersonalises the suffering. When so many are killed, the individual tragedies are somehow lost.

My old school keeps a register of every boy who passed through its gates from 1903 onwards. The register lists the house they were in, when they left the school and, where known, further details such as marriage, career and death.

The school had a strong academic reputation and parents queued up to spend much of their life savings giving their boys the best education they could afford. In January 1907, the register records that thirty six boys joined the school as bright young thirteen-year-olds. Some had brothers. Many were new to boarding school, perhaps waving tearful goodbyes to their mothers and fathers as the school gates closed behind them – much as I did a little over six decades later.

In the summer of 1911, a century ago and now passed into history, those boys left the school as young men, some heading for Oxford and Cambridge, others to Sandhurst where they would train to be officers. Confident young men, schooled to be leaders. Strong flag bearers for the school and their parents' pride and joy.

In less than a decade, the world was a different place. As war tore through Western Europe, the old order was swept away. And a generation and its values with it. Of those thirty six boys who had stood at the school gates a decade earlier, ten were dead, killed in the mechanised slaughter that was the Western front. A lost generation.

Three died on the Somme, young officers leading their troops across No Man's Land into a hail of machine gun fire.

Bright young things cut down in an instant. And three families more to receive those fateful telegrams from the War office.

For those of us who have never had to fight, their bravery is almost unimaginable, their idealism virtually inconceivable.

And so, at eleven o'clock on Armistice Day I stand in silence, this year as every other year. And I will certainly shake. Not because of Parkinson's you understand. But because, through the fog of the generations, I will imagine those ten. Not as the confident young men who fought, but as the anxious boys on their first day at boarding school. And I will think of my own son, the same age as those young men.

Pray God no more wars.

MENTORED BY A MADMAN

MENTORED BY A MADMAN is not your average scientific autobiography. But then Andrew Lees, its author, is not your average neuroscientist.

Trained as a physician at the Royal London Hospital in Whitechapel (as was James Parkinson himself), and as a neurologist at the Pitié-Salpêtrière Hospital (the domain of Jean Martin Charcot), he was director of the clinical neurology department at Queen Square from 1998 until his recent retirement. Not surprisingly, his more than seven hundred research papers and sixty review articles have made him the world's most widely cited Parkinson's researcher.

So, on the face of it, very much a mainstream, if glittering, academic career. A dusty academic perhaps, quietly treading the steps between laboratory and library?

Nothing could be further from the truth. Lees is a prolific writer with a string of books to his name, some predictable, others less so. Leave aside, for a moment, his books on Parkinson's (1982), Alzheimer's (2012) and tics (1985), all of which can be broadly considered within the direct trajectory of his career. Lees is a Scouser. Go back to the city of his birth and you see why his book on Ray Kennedy, the famous Liverpool midfielder who developed Parkinson's at thirty five, must have been a labour of love. Of course his fondness for Liverpool is broader still and The Hurricane Port, his psychosocial history of Liverpool is as vivid a portrayal as you will find of this most colourful of cities.

So what of this book, Mentored by a Madman?

The cover itself – black cloth, with imprinted bronze lettering and a quotation from the American writer Bill Burroughs – sets the tone. This is an elegant collection of memoirs from one of the UK's finest neurologist/neuroscientists over the last half-century. But there is so much more to the book than that. Certainly, Lees covers most of the key peaks in his academic career, and those peaks are very high, but that is not really what the book is about. This is a book about the relationship of hypothesis and experiment, of ideas and the means of testing them.

And this is where Bill Burroughs comes in. Burroughs is Lees's mentor, the madman of the title. Not in a direct face to face way – Burroughs and Lees never met – but in a very real way nonetheless, Burroughs was, for Lees, the teacher described by Henry Adams when he said "A great teacher affects eternity; he can never tell where his influence stops."

Burroughs's main influence on Lees was perhaps twofold – firstly to inspire a sense of life as personal adventure, and, secondly, to blur the boundaries between cultural and scientific endeavour, battle lines drawn by another scientist-writer, CP Snow, in The Two Cultures. And, in Andrew Lees, Burroughs may have found his ideal pupil. Lees can write. This book more than any he has previously written is testament to his ability to mix influences, to magpie pick ideas from disparate sources and to assemble all into a compelling literary whole. This book is less a statement of certainties and knowledge and more of ideas and hypotheses.

But what of Parkinson's disease? Is this a book about Parkinson's disease? Yes, in part, it is. But in many parts, that's almost irrelevant. The early use of L-dopa is described along with work on apomorphine, two drugs largely pioneered by Andrew Lees. Even if your interest is solely Parkinson's, there is plenty here to engage and interest. Lees tells a compelling story. A story of molecules and men.

At the same time Lees asks a lot of his readers. And if you want to get the most out of the book, you need to be up to speed on more than just The Naked Lunch. This is a book that will challenge your thinking, almost a scrapbook of ideas tumbling onto the page. Quotations and discussions of Bill Burroughs are everywhere. This is not just the world of William Burroughs – this is Kerouac, Ginsberg and Neal Cassidy. Lees may have written the first ever 'beat' memoir by a scientist!

If the book has a clear emphasis it is not upon the people as such but upon their interactions. Progress is not the points but the lines between them, the journeys not the destinations. And this book is full of wonderful journeys.

If I may end this review on an idea, it is an almost throwaway sentence in the chapter Looking for Clues where he talks about the complex physician-patient relationship and how critically diagnosis depends on knowing when to speak and when to listen. Patients were my main teachers, Lees says. And there it is in five words – the simple understanding that is the difference between being good at your chosen profession and being one of the greats.

WORDS

Sometimes I'm struck by the dissociation between the need to write and the ability to write. Some verbally constipated days I sit down at the keyboard with a clear idea of what I want to say but, try as I may, the words don't come. Or, on other days, they clatter onto the page like jackpot coins from a fruit machine, in riotous cacophony, tripping and tumbling. On some days the words run in measured meter beside me, jogging comfortably in my stride. On other days they weave and twist around me, sinuous musk-perfumed lovers, teasing and playing in the night. Sometimes words are bold, brazen, sassy and obvious. Funky words, big bass words, power chords of twang kerrang. Sometimes they are evening larks, shimmying in the sunset.

Words are me and words are what I do. They are my daily bread and my nightly jam. They are prescriptive, descriptive, seductive and deductive. They find a path through the wilderness. They lead me into trouble. They are my best friends and my worst critics. They are a source of confusion and a spangle of clarity. They are life. They are death and all stations in between.

Sometimes, like the currency they are, I am thrifty, spending each reluctantly, clicking shut the purse. Other times, like a lottery winner, I stand and throw notes to the wind. Sometimes the words are static, lumpen globs on the page, like my granny's butter bean dip. Sometimes they splash like Chablis, cascading in tiny droplets round the world.

Words lift me when I'm sad, comfort me when I'm bad, sing with me when I'm glad and taunt me when I'm mad. Words climb mountains in the snow and are warm sand between my toes. They are the flickering tallow by my bedside and the lantern in the porch. They are the ghost candles in the cemetery and the hand of glory.

They are the threads of life, the tapestry of time, the voice of my friends and the whispers of lovers. They were the first things I heard and will be the last things I utter. They are the tortured path through my desktop's clutter.

A MESSAGE FROM MR GRUMPY

I have Parkinson's and I work for a Parkinson's charity. During the working week, I eat, sleep, breathe and think Parkinson's. It's unavoidable. It's part of my makeup, whether I like it or not. It's the first thing I think of when I wake up and the last thing I remember before I go to sleep. It's part of what drives me on. And it's full on.

But when the weekend comes, I try to forget it. I try to close the door on Parkinson's. I like to see my family and friends, watch cricket, listen to music, enjoy a glass of wine, take photographs, make glass, think and write. Forty eight precious hours each week in which to clear my head. And every hour is precious.

The weekends are a Parkinson's free zone as far as I'm concerned. And as much as is possible. So if you want me to think about Parkinson's at the weekend, remember what you are taking away from me. If I sometimes seem a trifle grumpy, that may be the reason why. Or I may just be a grumpy old git.

HUNGER PANGS

I've gained a bit of weight recently. Well, more than a bit. In fact I have gained so much weight over the last year that few of my clothes still fit. Many can only be worn in circumstances where blood supply to the lower limbs is not essential.

My son, ever the comedian, walks past me and pretends to be drawn in by the gravitational field of my stomach. Okay it was funny once. Still, it amuses him. In any case, as I frequently point out to him, I was not always thus. I was as slim as a rake at fifteen. My rib cage would have made a quite passable xylophone. Visitors to our home would speculate on why my parents were apparently starving me.

But fast forward four decades or so and it's a different story. I am, in my own context, the size of a barrage balloon. In the decade since I was diagnosed with Parkinson's, I have gained approximately four stones.

Now my weight, like everybody's, is a balance between caloric input and energy expenditure. In plain terms, this means how much you eat and how much you exercise. It's not rocket science – every extra doughnut means another half-hour on the exercise bike. Or whatever.

So where do I lay the blame for this precipitous increase in weight? Obviously diet and exercise are the first places to look. Certainly I don't get anywhere near as much exercise as I would like or as I recognise would be beneficial. I don't know whether I make excuses too readily or if my day is simply fuller than will accommodate. Either way it doesn't seem to happen.

So what about diet? Well I do have a certain penchant for Jaffa cakes (which are, incidentally, a cake and not a biscuit despite what others may tell you) and soft French cheeses. And most of the time, when I am working on my own, I may tend to nibble. On the whole however, I forget mealtimes. Many's the time that I will reach five in the afternoon and realise that I have eaten neither breakfast nor lunch. On other occasions I may gorge on leftover pizza for breakfast, fast food for lunch and the odd Murray mint in the afternoon.

My point – and you'll be glad to know that not only is there one but I have finally reached it – is this. People with Parkinson's (PWPs) don't always know when they should eat and when they should not. In the absence of external cues, such as the entire office decamping to the Pret a Manger, we do not seem to recognise physiological hunger and satiety cues in the same way as people without Parkinson's. In other words, we don't seem to realise when we're hungry and when we are full.

In some ways this should be no surprise. There is a gathering school of thought that Parkinson's starts in the gut nerves before progressing to the brain areas that control movement. That would fit. If those nerves that tell us it's time for a sandwich or that we don't need a second helping of raspberry ripple are damaged, our frame of reference goes with it.

This can probably go one of two ways. Either one simply forgets to eat or one nibbles most of the time. Obviously it's clear which camp I occupy but for every one of me, there is another human skeleton who has simply forgotten to eat. And if you live on your own, without the social reminders to eat at predefined times, the problem is exacerbated.

I feel that this is probably an underresearched area. We know that diet can influence the absorption and effectiveness of medication. So shouldn't we be trying to find out more about this and how it may influence day-to-day management of the condition?

Just a thought.

2017

INTERESTING TIMES

We live in interesting times. In a little over a week since his inauguration, President Trump has instigated perhaps the most isolationist agenda ever conceived by a world power. By a mixture of legislation and construction, President Trump seems determined to take America to some kind of modern day Xanadu where it will be safe from the big bad world behind its walls and impenetrable xenophobic legislation. Apparently this will make America great again.

Wrong. Totally wrong.

The construction of a physical wall between the United States and Mexico in order to limit illegal Mexican immigration is proffered as a social measure – an effort to protect American jobs from immigrants working illegally. But of course the replacement of economically invisible Mexicans by American workers paying American taxes provides a more potent legislative stimulus and presumably is the rationale that underpins the legislation. The translation of this economic concept into a physical wall is somehow both tragic and comic - tragic that any adult can believe in its effectiveness and comic in its misplaced logic. The majority of illegal immigrants enter via airports and conventional border crossings. In any case, many enter the US legally, only becoming illegal when they fail to leave.

The decision to build the wall is antagonistic and stupid. It has riled not only Mexico itself but more or less every Hispanic living in the United States.

But the wall, magnificently absurd though it is, pales into insignificance next to the travel ban imposed on refugees and residents of a range of Islamic countries. This appallingly misplaced legislation serves no useful purpose and runs counter to all logic. Countries have traditionally benefited from refugees bringing new ideas and challenging the status quo. Whatever happened to ""Give me your tired, your poor, your huddled masses yearning to breathe free"

In the space of two asinine pieces of legislation, President Trump has antagonised a major ethnic group within his own borders and virtually declared war on Islam.

This is not how you make friends and influence people. This is not how countries interact. Let us not delude ourselves that Trump's attitude is exciting and fresh. It is cringingly embarrassing. This is not how presidents of world powers behave. And far from being fresh air in the corridors of the White House, this comes across as tired bombast from a man who simply falls short of the job requirements. Mr Trump is an international laughing stock.

Worse than Mr Trump's standing is the message from history. The wave of state endorsed anti-Islamic feeling we seem to be seeing in the United States has a chilling historical precedent in the treatment of Jews in Germany during the 1930s when economic and political persecution came to a head on the night of the ninth of November 1938. Buoyed by populist, state-supported racism, the people of Germany destroyed around eight thousand synagogues and Jewish businesses in what came to be known as Kristallnacht.

We would be wrong to delude ourselves into believing that this could only happen somewhere else and at another time. In many respects, the seeds are already sown by the sentiments encapsulated in this legislation. America is no longer a country where Muslims can feel welcome. It is barely a country where they can feel safe. And in some places it is not even that.

The ethical arguments against these new pieces of legislation are strong. But even if we put aside these ethical arguments, the ultimate destination of the legislation is against the United States best interests. Withdrawing behind one's borders and hunkering down is not the way to secure American influence on the world stage. Much of the security of the world over the last half century has been predicated on a balance between two world superpowers. If we are reduced to a single world superpower – Russia – the world's a much more unsettling place. America's reluctance to endorse NATO further supports the view that the US is abrogating its responsibility as a world superpower. Retiring behind its own front door and closing the curtains will not make America great again. President Trump is wrong. He needs to rethink.

To quote Oliver Cromwell "I beseech you in the bowels of Christ – think it possible you may be mistaken".

IF YOU HAVEN'T GOT IT, YOU DON'T GET IT

Viewed as mediator between the patient and their pathology, the role of physician is an unenviable one, constantly balancing the sometimes conflicting requirements of clinical judgement with empathy for his patients. Nowhere is this balance more tricky than in general practice where physicians often have lifelong relationships with individuals and families. But does this familiarity help? The argument goes something like this: The physician must establish a firm and accurate picture of symptoms, signs and treatment options. It is also anticipated that, as much as possible, they wield this knowledge with understanding and empathy for the patient's plight. Of course, empathy comes at a price. The more doctors put themselves in their patient's shoes, the less emotionally detached they become. Once detachment is compromised, judgement soon follows. Impaired judgement translates into less rational decision-making which, in turn, inevitably means poorer care. Taking the argument to its logical extreme, empathy is bad, an obstruction to best medical practice.

This model, if model is the word, certainly held sway a generation ago. My father, a Yorkshire GP for some forty years, subscribed to a rather detached approach based on sound clinical judgement. He was a fine diagnostician in his own patrician way. His practice partner was an entirely different kettle of fish. What he lacked in clinical skills, he made up with attention to wider patient concerns, bandying about words like "holistic" and "dialogue", terms unheard of in 1960s Yorkshire. The practice receptionist would greet each patient through the surgery door with the simple, if unsettling, enquiry "have you come for a cup of tea and a chat or do you want to know what's wrong with you?" The implication was that you could have the one but not the other.

This of course is tosh.

I would go further and say that not only are best clinical judgement and empathy with your patients compatible with each other, but that they are necessary components of genuine patient-centred healthcare. The issue is how this can best be achieved. The first part is easy – just pay attention in medical school, take notes, ask questions, offer opinions and learn from your mistakes. Easy.

The second part – empathy with patients – is a bit more of a challenge. I learned it the hard way.

For a couple of decades at the tail end of the last millennium, I was a researcher and academic at what was then the London Hospital Medical College in Whitechapel, attached to the Royal London Hospital. My main interest was in the function of monoamines such as serotonin and dopamine in the brain. Inevitably it wasn't long before I was drawn to the basal ganglia, where I concentrated on the function and control of dopamine in Parkinson's disease. In all, I spent twenty three years at "The London" rising from PhD student to Reader before leaving academia in 2003. Three years later, in one of those coincidences in life that you can do without, I was diagnosed with Parkinson's.

Despite the fact that I had lectured on Parkinson's to a generation of medical students, I still somehow managed to persuade myself for a year or so that the symptoms I had amounted to something else. Somehow I did not connect the way I was feeling with the symptoms I was exhibiting. It is perhaps a bellwether of the isolation of academic research that the first actual patient I met with Parkinson's was myself.

So what does it feel like to become a patient with the same condition you have researched and taught for more than twenty years?

For me it was initially a challenge to marry up the externally observable (objective) manifestations of the condition with the very subjective experience. But this difficulty inevitably spawned a wider interest in the dynamics of patient experience and physician assessment. Parkinson's is a good test bed for this – the physical manifestations of the illness are relatively straightforward for the physician to understand, comprising purely motor symptoms. Yet the full orbit of the condition is only apparent with detailed interrogation of the patient experience, thus unmasking the many non-motor symptoms that determine quality of life for the patient. These, when inadequately appreciated by the physician, create a false clinical picture of the condition and lead to poor decisions on management. It is a phrase oft repeated among the Parkinson's community that "physicians know what Parkinson's looks like but only patients know what it feels like". Put even more succinctly "if you haven't got it, you don't get it".

Taking that philosophy to its logical end point, the people best equipped to describe Parkinson's and to influence its treatment and management are of course physicians who have the condition itself. Although patients can and do eloquently describe the personal burden of Parkinson's, their words hold little sway with the medical community. Physicians listen to other physicians. That is the way of things. And if we, as patients, want our opinions to be taken seriously, this small band of patient-physicians represents a powerful conduit for transduction of the patient experience into clinical influence.

Of course Parkinson's is just an example. The same principles apply to most if not all other conditions. If we want to know what is important about an illness, we need to turn it inside out.

RETIREMENT BECKONS

For the last thirty seven years, I have worked in the field of Parkinson's in one way or another. First as a PhD student studying dopamine release in the basal ganglia, then as an academic leading a small research team and, more recently, working as a patient advocate with the Cure Parkinson's Trust. I have had Parkinson's myself for 10 years.

So, all in all, I think I've pretty much paid my dues in the field of Parkinson's. And as I've grown older, I find work harder and Parkinson's harder still.

Parkinson's takes its toll in so many ways but especially on family and friends. And I'm aware that the harder and longer I work, the less time I have for my family.

Weighing everything up, I made the decision recently to take early retirement, perhaps appropriately, at the end of April, Parkinson's Awareness Month. It's time to take a back seat and to have more time to reflect and focus on a handful of smaller but stimulating initiatives. I'm not giving up on the Parkinson's – just channelling my energies into more focused areas. And above all, I'm not giving up on the many friends I've met through Parkinson's. You guys have kept me going and I hope you will still.

Retirement will give me time to think, to write and to do all those little jobs I've been promising to address. But, more than any of this, it will give me the chance and time to try to be the father I should have been all along.

WHO IS THIS IMPOSTER?

If you open almost any article on the Internet about James Parkinson, the physician after whom Parkinson's disease is named, you will see the same picture. A middle-aged man with beard and moustache. But if you're trying to trace the picture back to its source, things get a little more complicated.

This may well be a picture of *a* James Parkinson but it is emphatically not a picture of *the* James Parkinson.

How do we know this? Very simple really. All we have to do is look at the bare facts of James Parkinson's life and that of the history of photography.

James Parkinson was born on April 11, 1755 in Shoreditch. His most famous work "An Essay on the Shaking Palsy" was published in 1817. But most importantly for this issue, James Parkinson died on December 21, 1824.

So what, I hear you cry.

The problem is quite simple. The very first recorded photograph was taken by Joseph Nicephore Niepce at Le Gras in 1826. That's two years after James Parkinson died. The first photographs in Britain were taken by William Henry Fox Talbot in the 1840s.

It's probably impossible to say with confidence who the bearded man is. But we can say with confidence who it isn't.

Parkinson's disease presents many challenges. It's not always easy to identify the condition in new patients. It also appears to be difficult to identify the man himself.

200 YEARS TOO LONG

Each year at about this time, we talk about Parkinson's Awareness Day, Week, Month or what have you. And each year we hear the usual talking heads telling us that lots is already being done to make the Parkie world a happy place. We should pipe down and be grateful.

That's drivel. Complacent drivel.

That's why I am slightly puzzled by the commemoration of the Parkinson's Disease Bicentennial. In case you are unaware (and if you are, where have you been hiding?), let's be clear what exactly that is. Parkinson's disease as we now know it was first described by James Parkinson in 1817 in his "Essay on the Shaking Palsy" although it didn't acquire the Parkinson's moniker until some years later when Jean Martin Charcot generously ascribed pre-eminence on the condition to Parkinson.

Today especially, throughout the globe, neurological associations, patient groups, and charities are falling over themselves with invited lectures, symposia, conferences, monographs, webinars and publications marking this bicentennial. The whole jamboree has the air of a celebration.

But what precisely are we celebrating? Surely not the fact that this condition has been recognised for two centuries yet still has no cure. Indeed, until fifty years ago, there were barely any symptomatic treatments.

To my mind, this is less a cause for celebration than for collective mourning. We, the Parkinson's community, should not be proud that a neurological condition has been known for two hundred years yet has no cure. This is a damning indictment not a ringing endorsement.

That we have no cure is a collective failure. It's all too easy to place responsibility at the feet of the research scientists or of the drug companies. But this would be wrong. We patients must bear our share of the responsibility. Clinical trials fail for a number of reasons. Sometimes poor experimental design is responsible. Sometimes inappropriate statistical analyses are to blame.

But as often as not, trials fail because of inadequate recruiting. And that one is down to us, the patients.

Our failure to find a cure – and let's be clear, this is a failure – is the collective responsibility of the entire Parkinson's community, from doctor to patient and from scientist to caregiver. We have all failed in some way to advance the field as far as we should have.

We have failed also to raise awareness of Parkinson's in political and governmental circles. Hansard has recorded all the business of the Houses of Parliament for the last 200 years and, during that time, Parkinson's as a health issue has not been discussed once in the Commons until this year.

In March this year, Nick Thomas-Symonds, MP for Torfaen, raised the issue of Young Onset Parkinson's Disease in an adjournment debate in the House of Commons, largely at the instigation of Spotlight YOPD, a charity founded to cater for the interests of younger Parkinson's patients who feel largely neglected by the country's major Parkinson's charity. It's hard to believe that this year is the first in two hundred that Parkinson's has been discussed at a parliamentary level. Hard to believe but sadly true.

We need to work harder to cure this illness. Parkinson's will not go away until the entire community from caregivers and partners to pharmaceutical multinationals and research scientists pull together. Everybody has a part to play and, until we all play those parts, we will not see an end to this condition.

11 April 2017 is a day when scientists, physicians and drug companies will doubtless congratulate themselves on the progress they have made over the last two centuries. And, don't get me wrong, we have made progress. But the truth is that we have not made the progress we want and need. Every year around a quarter of a million people with Parkinson's die. And they die because we don't have a cure

for this illness. So 11 April 2017 is not a day to celebrate Parkinson's disease. It's a day for us to look at ourselves in the mirror, reflect on the human calamity of Parkinson's and to renew our vows to beat this illness once and for all.

EXTREME GARDENING

Gardening is not my thing. And when I say 'not my thing' I mean not my thing in the sense that large wooden stakes driven through the heart are not Dracula's thing. Or Christmas is not your average turkey's thing. That sort of 'not my thing'.

On either side, my neighbours cultivate beautiful patches, manicured lawns and chocolate box flowerbeds. Hanging baskets positively drip with geraniums. Pansies tessellate along their borders. A riot of colour and perfume assaulting the senses.

Separating these horticultural nirvanas is my humble dwelling, a rotten tooth between two pearly whites. Weeds claw their way through the paving of the driveway, gently gnawing away at the house's foundations. In my neighbours' gardens, nature is controlled, brought to order and made to serve. In my garden, nature is in control, gradually and in a thousand tiny ways, taking back its own. My house is nature's revenge for my neighbours. Here nature dictates. If my neighbours' gardens are benign nature, mine is a more malignant. My garden is Turner to their Constable. Nature in its pomp rather than in retreat.

Leaving aside the not inconsiderable matter of being ostracised by the neighbours, and I'm keen to avoid that, there is the issue of fighting the weeds in my own way. I need practical and simple solutions to gardening issues.

Take the lawn for instance. I simply don't have the strength to wrestle a lawnmower over the garden every weekend during the summer. Not that it would matter if I did. Even if I did have the strength, I certainly don't have the interest. Let's say that it takes say an hour to mow the lawn, tidy up and trim the edges (and at Parkie pace that's probably an underestimate). Obviously the lawn doesn't need trimming during the winter but let's say that's still thirty-five weeks out of fifty-two when it does. Let me speculate also that I will live for another fifteen years perhaps (I'm coming up to sixty now) and will continue to mow the lawn. By the end of my life, I will spend five hundred and twenty-five hours mowing the lawn. Assuming that one doesn't mow the lawn in the dark, and who am I to say what you keen gardeners actually do in the privacy of your patches, that amounts to forty-four days of lawn mowing. A month and a half.

Now I don't know about you, but I can't help feeling that when I get to the end of my life, I would probably quite like to have those forty-four days back.

The solution of course is simple. If you want the look of grass without the weekly faffing about with tools, artificial grass makes a lot of sense. I reached that conclusion a year ago. And for three days in August last year, a chirpy team of Latvians removed my lawn and replaced it with an artificial surface.

Does it look artificial? No, not in the least. Visitors have constantly been surprised at how persuasive it is. Gone are the days of Astroturf, which looked blue rather than green, more of a sports then recreational surface. No, modern artificial grass is realistic. But then why wouldn't it be? Even my neighbours, grudgingly perhaps, conceded that it was a fine replacement for the lawn. And I have my forty-four days back.

Of course weeds are less amenable to this kind of solution. My local garden centre offers a number of solutions – mostly involving chemicals that will poison the weeds. Sometimes involving chemicals that will do this without also poisoning your own pets or the neighbours' cats. But for the most part, these are slow acting, fiddly and unrewarding. I don't want to poison the weeds and then, over several weeks, gloat over their misfortune. No I want a swift and decisive solution. I want the kind of solution that will make a weed think twice before poking itself out of my driveway. A short sharp shock.

And I think I have found it.

Once again, Amazon has come to the rescue. Nestling among the more reclusive pages of the gardening equipment section lie a handful of entries devoted to alternative ways of controlling weeds. To cut a long story short, these "alternative ways" are fundamentally flamethrowers.

Now that is more my kind of gardening. So inevitably I ordered one. That's right, I ordered a flamethrower.

I order a lot of things from Amazon but I have to say I have not looked forward to the arrival of an Amazon package quite so much for a long time. It's due on Tuesday and I am practically counting down the hours. Only five more sleeps.

But my excitement is tempered by an almost equal sense of foreboding among my children. They have each, in their own ways, pointed out to me that a man with Parkinson's in control of flamethrower is not a brilliant combination. My eldest suggested I look through the fine print on the house insurance before picking up the flamethrower. Pah! They worry when I play with scissors.

But I ask you, seriously, what can possibly go wrong?

PUBLISH AND BE DAMNED

One of the most important aspects of any piece of research is the publication of the results. This is the point where you submit your work for scrutiny by your peers. That in itself is always a heart-in-mouth moment when you discover what they really think about your research!

But if you think that's difficult, just try one of the electronic manuscript submission programs so popular among the journals these days.

When I were a lad it were simple (last sentence to be said in a kind of maudlin Yorkshire accent).

You simply bundled three copies of the manuscript into a big envelope and winged it on its way to Brain Research, Journal of Neuroscience or wherever. Something like three months would elapse and then you would receive either a thin or a thick envelope. The thick envelope usually signified that the reviewers wanted massive revision to your work before they would even entertain the idea of looking at it again. A thin envelope could mean one of two things. Either your manuscript was so awful that the reviewers could not bring themselves even to look at it. Or the manuscript had so completely dazzled the reviewers that they were literally speechless.

Over the years that I was an active researcher, I saw pretty much all of these types of envelope. On one occasion, in a scene reminiscent of the recent Oscars debacle, I received two envelopes on the same day – one rejecting the manuscript and the other accepting it. A couple of days later the third envelope arrived to tell me which of the first two was the correct assessment of my work. Fortunately it turned out to be the more favourable although I've always suspected the decision was made by their legal department, fearing litigation from a thwarted author.

Nowadays the process is much more streamlined. At least from the point of view of the editorial offices. No longer do they receive acres of rainforest each day. No longer does their shredder burn red at night. Instead, the editorial office is a buzz of electrons tripping their way to and fro. In the modern digital age you can reject an author's work in milliseconds. No longer must those manuscripts return to the sender, like Phileas Fogg, by rail, steam and road. The press of a button and a clatter of electrons commits the author's life work to an electronic dustbin.

But I'm jumping ahead. Already we are talking about assessment and rejection of the paper. Let's go back to the beginning and the simple process of submitting a manuscript.

Digital manuscript submission is massively time-saving for the editorial office. It achieves this by being massively labour-intensive for the authors submitting the paper. In days of yore, a covering letter went with the manuscript (so they would know where to send the rejection letter) asking them to consider it for publication. That was it. No email addresses (not that there were any when I started submitting papers), Twitter handles or web pages. All the spade work was done by the editorial office who had to allocate reviewers, despatch the article to them, wait for the reviews, collate them and distil the thoughts for the hapless author.

Now it's different. Writing the manuscript is a doddle compared with the process of submission. Usually this is a sequence of maybe a half-dozen sequential screens. Each screen must be completed in sequence and may contain up to twenty different elements. Nearly all will have that tell-tale asterisk, indicating that this is a "mandatory field". One which must be completed – and I mean completed – before progression to the next level. Prince of Persia was a walk in the park compared with submission of a scientific manuscript.

The first screen usually has the basics – your name, address, email, office telephone number, mobile telephone number, inside leg measurement, names of children, bank account details, political leanings, credit card numbers, blood group, sexual preference, and keys to your car. Assuming you fill that out correctly, you may move onto the next level. Of course this won't happen the first time. No, on the first time you will receive a message telling you that you have inappropriately capitalised one of the words. Needless to say, it won't tell you which word. That's the first part of your initiative test. So you spend twenty minutes, changing each response, one by one. If you're lucky, it will retain your existing answers. If you're unlucky, each will have to be re-entered from scratch. And if you're really unlucky, you will simply get a message saying something like "syntax error in line 28". You almost certainly don't know which is line 28. In any case it doesn't matter. All the program is really saying is "you are an idiot who cannot fill out a form correctly. I blow my nose at you."

By the end of the first hour you have completed perhaps two of six screens, sworn at your co-workers and downed your third espresso of the morning. By the end of screen four, the journal knows more about you than the CIA and on every street corner there seem to be men in dark glasses with walkie-talkies. But often, it never gets that far. By the end of the entire process, the screen asking who should pay for the wall across the Mexican border, you have lost the will to live and are weighing up different exit options. Or at the very least burying your keyboard in the computer screen. Or deciding that you didn't want to be a scientist anyway.

You click submit and immediately an email pops up in your inbox telling you that there will be a processing fee for your manuscript. You pay $1000 or something like that for the privilege of having your work assessed. Unbelievable. Even more unbelievable is the fact that many authors accede to this. It's rather like having to buy your own Christmas presents. But nonetheless, you agree. Then another email arrives saying there will be page fees and additional costs for the reproduction of colour figures. And that's in an online journal.

I can't help feeling, and I say this as both an author and a member of several journal editorial boards – a sort of poacher turned gamekeeper, that the process really doesn't need to be this difficult. I would be very interested to know which journals receive the most manuscripts. I can't help feeling that it will be those who submission process is easiest. The ones which don't need the complete addresses and emails of every single one of the thirty authors on the paper. The ones which don't require electronic signatures. The ones which don't charge for every bell and whistle that your paper doesn't need anyway.

Everything we do that makes the job of the editorial board simpler makes the job of the scientists more difficult. Aren't we getting this the wrong way round?

HOW TO SURVIVE RETIREMENT

When I was a child, my enduring impression of retirement was one of paralysing boredom. I saw ancient wrinkled creatures who, having earned their retirement through years of hard graft, simply had no idea what to do with the time now available to them. Their daily habits, forged through years of employment-based routine, could not adapt to the wide-open vistas of retirement. Having lived with days filled to the brim by the needs of others, they did not know what to do with the long days available for their own pleasure.

And it was often no better for spouses. Used to their husbands leaving for work at 8:30 and returning at 5:30, the forlorn figure in the living room, sighing over a cup of tea and a crossword puzzle, was equally alien. The conversation over dinner time "how was your day?" was replaced by nothing. All the little rhythms of life in employment were replaced by the slow tick tock of the engraved carriage clock on the mantelpiece, and its solemn chiming reminder of the years that sped by in a flash in employment and the hours that will not budge in retirement.

It's no wonder that retirement kills. Men and women who have spent their lives at a desk, bedside, factory, office, field or barn are ill-equipped to deal with a life devoted to themselves. It's the same with prisoners, so imprinted with the life and rhythms of incarceration that they are often unable to understand their liberty.

For men, and I think it's primarily men, a job is a defining feature and a reason why unemployment casts such a long shadow. Men without employment struggle with their identity. A man who has been a bank manager for instance finds the loss of status hard to stomach. A surgeon, used to barking orders, cannot easily live when life no longer listens.

Retirement, in many ways, is unemployment painted larger. Unemployment tells you that you no longer have a job. Retirement tells you that you will never again have a job. But worst of all, for men who thrive on the status that their position afforded, is the knowledge that one is no longer needed. That one is replaceable.

Retirement then is not easy. Many companies recognise this and invite employees approaching retirement to attend classes that will help them prepare for this major life change. And they do well to do so.

Me, I've been retired for two weeks now and already I can sense the pace of life changing. Activities, rather than being abbreviated by time now expand to fill time. The rhythm of life is different. I hear the birds in the bushes, I notice when flowers bloom. The coffee smells better and the orange juice is sweeter. Yet nothing in actual fact has changed, merely my appreciation of it.

I have a theory about how to survive retirement. It's probably not original. But it is simple.

You need hobbies. Specifically, you need two hobbies – one indoor and the other outdoor. And no, doing the crossword does not count. Before you all write in to complain, remember this is my theory.

An indoor hobby will sustain you in autumn and winter and in the evenings while an outdoor hobby is perfect for spring and summer. Of course two hobbies is merely the minimum. Three, four or a dozen is fine as well. But two seem to balance each other.

I have no science to back this up nor reading to support its scholarship. And of course there are exceptions to this rule. There are plenty for instance who contend that golf is the route to all human happiness in retirement. I tend to side with Mark Twain on this one.

My theory clearly needs research. Painstaking research over many years. And I would like to assure readers that I'm prepared to dedicate myself to this task for as long as I'm able. I'm prepared to stay retired and to enjoy myself. Yes, I will push myself that extra mile. All in the name of science you understand.

THE LIGHT THAT BURNS TWICE AS BRIGHT...

Let's start with the basics. Tom Isaacs, president of The Cure Parkinson's Trust and co-founder of Parkinson's Movement died unexpectedly a couple of weeks ago. For CPT and PM, the loss is particularly acutely felt. Over the last many days we have gradually informed people of this sad news. The reaction has been astonishing. Once over the initial shock of the news, people have been at pains to tell us how much Tom meant to them. At CPT, we knew Tom was well liked in the Parkinson's community but we had no idea how much further than that it went.

Somehow Tom seemed to reach out to people, to make them feel part of things, to involve them. Over the last week people have told us story after story of Tom and the ways in which he had made them feel special, important and valuable. He had a way, aided by Helen, his right-hand woman, of making your issues his issues, of making your concerns his concerns.

In anybody else but Tom, this might seem contrived or calculating. But there was nothing contrived or calculating about Tom. If he felt you were wrong about something, he would find a way of telling you that would let you rethink the problem yourself. He could be direct too. Especially if you knew you well. It wasn't rudeness or anything like it. It was simply a reflection of the fact that time is precious to us Parkies and that the sooner we could get to the point the better. The same directness led to a great deal of banter and verbal sparring. He and I loved taking the Mickey out of each other. Tom, like me, had no time for the political correctness of PWP (people with Parkinson's). For both of us, we were Parkies, plain and simple.

His sense of fun spilt over into his talks and presentations. Although often simultaneously trying to deliver an important point, he loved telling jokes, largely at his own expense. And his audiences loved him. There was simply no better after dinner speaker than Tom. He would have diners in stitches with jokes about dyskinesias and auction rooms, or whatever took his fancy. Even when the audience had heard the joke before, he would thank them for laughing again.

Inevitably, his talks overran. Over the years I have spoken at many CPT meetings and learnt, through experience, that being given the timeslot immediately after Tom is the equivalent of a hospital pass in rugby. Whether humble students or feted professors, it mattered not one jot – Tom upstaged them all. He took particular pleasure in rattling the cages of any scientists or neurologists present. And yet, this teasing was delivered with such a twinkle in the eye that it was impossible to take offence.

He had an uncanny knack of making a serious point with a joke. He disarmed audiences with his sense of humour making people listen where otherwise they might not. To be able to speak about stem cells at a conference in the Vatican is remarkable enough. But incredibly his combination of hard science and an infectious style of delivery led to a personal audience with his Holiness Pope Francis.

But it was at the WPC meetings where he was most in his element. Tom needed a big stage, and not just because of his irrepressible dyskinesias that would send him careering across the platform with the audience holding their breath as he lurched towards the edge. He had big ideas and needed space to spread them out for an audience. He often spoke in sessions with eminent scientists and physicians. But he was never overawed. And, more than anything, he was never a token patient. Tokenism was anathema to him. If the patient voice was to be heard, it was to be heard properly and on an equal footing to that of the clinicians. Tom was, in Parkinson's terms, a rock star.

But what is left behind and how shall we best endorse his legacy?

Before Tom, no serious scientist or clinician would use the word "cure" in the same sentence as the word "Parkinson's". The notion that Parkinson's disease was incurable was ingrained in every medical student from the outset. Tom simply asked the question "why?" And the more he asked the question, the less satisfactory were the answers. But more than that, Tom made physicians start to ask the same question – why is Parkinson's incurable? Now we talk openly of a cure. Certainly we're not there yet but we are on the way. It is only a matter of time. By asking that simple question, Tom has induced a sea change in our thinking about Parkinson's which, in turn, has influenced the direction of clinical research. Not bad for a little man with a handful of O-levels.

Tom was often exasperated by the pace of clinical research, or more accurately the lack of pace. Again, he found a simple solution. If there was not enough research taking place, he would find a way of funding more by starting a research charity – The Cure Parkinson's Trust. A little over a decade old, the charity is the embodiment of Tom's ethos.

Tom felt very strongly that the patients were the key to a cure for Parkinson's. He felt that patients were, in the classical patrician model of healthcare, an underutilised resource. He believed that it was the interface between patient and physician that would yield the answers needed. He believed in collaborative medicine and he believed in a powerful, strong and unified patient voice. His leadership helped to make patient advocacy a vital component of drug development. Before Tom there was no such thing as a patient opinion leader. Now there are many.

Above all, Tom stood for one thing – friendship. He believed that friendship, one-to-one, in groups, and in communities was the route by which everything could be achieved. He believed that friendship and teamwork were not the best way to achieve results so much as the only way.

No reminiscence of Tom would be complete without mention of his singing. Never was there a more joyous, uninhibited and funny singer. He had us all in stitches with his alternate takes on classic songs from the musicals, most immortalised on Youtube.

Tom was a good friend, a great leader and a true inspiration. If only he had lived longer. But, in the words of the Tao Te Ching:

"The flame that burns twice as bright burns half as long".

EXENATIDE – REAL HOPE OR ROYAL HYPE

I've seen enough false dawns to take a jaded view of new breakthroughs in Parkinson's – and with good reason.

People with Parkinson's have lived with the same tired handful of medications for too long; our best drug therapy, levodopa, is fifty years old. Apart from a handful of other drugs (dopamine agonists or inhibitors of dopamine metabolism), that's your lot. These are drugs to treat the symptoms but nothing that will slow down the progression of the underlying illness. Such a thing would be perceived as the holy grail of Parkinson's therapy.

Treating the symptoms is all well and good but eventually the drugs can no longer mask the underlying neurodegeneration. At that point your treatment options begin to dry up.

Until now that is. Friday's publication in The Lancet from Prof Tom Foltynie's group at University College London [1] examined the effect of once weekly Exenatide, a drug derived from the saliva of the Gila monster (I'm not making this up) on motor symptoms in Parkinson's patients. They found a substantial and significant difference that outlasted the period of treatment by three months.

This finding is remarkable for two reasons. Firstly, it reports the first data indicating that the speed of progression of Parkinson's can be slowed. The significance of which cannot be overstated. Nothing up to this point (and we can debate the Rasagiline data until the end of time) has shown evidence for disease modification. The paper's authors are cautious of making such a claim; the data is nonetheless exciting.

Of course, we are used to media hype, but this is something quite different. This is a placebo-controlled clinical trial of a drug in humans. This is data with direct relevance to the wider patient community. We do not need to extrapolate. We do not need to hype the hope. The results are clear-cut. I am as jaded a scientist as you get. I've seen it all before. But I'm persuaded by this new dataset. I believe this is probably the first evidence of disease modification.

The second reason why this was a remarkable study is, in many respects, just as exciting. This is not the trial of a code number drug XYZ 123 at the early stage of the development cycle that may last ten to fifteen years. Exenatide is a medicine that is already marketed. It is a drug for diabetes that has been repurposed for Parkinson's.

Why is this important? It is important because we already know the medication is, in broad terms, safe. (I should at this point probably insert a long caveat about the meaning of 'safe' because, ultimately, nothing is absolutely safe.) Much is known about the toxicity or lack of toxicity of the drug. In essence, we can lop years off the development cycle of a new medication by using an existing one. The drug is available clinically, albeit for a different purpose, and, in theory at least, you or I could take the drug tomorrow.

So what is there to stop us? Why should we not visit our local neurologist and persuade him to prescribe it for us? Patients are understandably impatient when it comes to new medications. Many may well try to take this foreshortened path to improve their own health. In some ways it's hard to counsel against this. With any degenerative condition, time is not on our side but we should nonetheless be cautious. We cannot be certain that the drug will behave the same way in people with PD as it does in people with diabetes.

So let's be a little cautious... at the same time let's rejoice in these findings. This is genuine cause for excitement – and it's long overdue.

[1] Dilan Athauda, MRCP, Kate Maclagan, PhD, Simon S Skene, PhD, et al (2017) Exenatide once weekly versus placebo in Parkinson's disease: a randomised, double-blind, placebo-controlled trial. The Lancet DOI: http://dx.doi.org/10.1016/S0140-6736(17)31585-4. Published 03 August 2017.

3AM IN MICHIGAN

It's 8 AM. Which is fine if I was where the 8 AM was, if you see what I mean. Kent, in the south-east of England. But I am in Grand Rapids Michigan where the time is 3 AM. And here is where the problem lies. I may physically be in Michigan but my sense of time is firmly on the other side of the Atlantic. As far as my sense of time is concerned, it's a case of wakey wakey rather than sleep.

So why am I in Grand Rapids, that quirky jewel in the Michigan landscape?

It's a fair question and it seems even fairer at 3 AM when you begin to question pretty much everything.

I, along with a dozen others, are part of the CPT (Cure Parkinson's Trust) party here for the Grand Challenges in Parkinson's Disease and Rallying to the Challenge meetings on Parkinson's Disease at the Van Andel Institute. The first is primarily a scientific meeting, focused on the latest advances in our understanding of the condition, while the second is targeted mainly at specific issues affecting people with Parkinson's. But both occur concurrently and there is some overlap between the two.

But aren't you retired from all that stuff, I hear you ask. Well, yes, I answer sheepishly. It's true that I did retire at the end of April from the CPT after six years working for them. I hadn't anticipated attending the meeting if I'm honest. But a lot has happened at CPT since I left, not least the passing of Tom. Tom, along with Helen, was the spiritual heartbeat of CPT, its central ethos. His unexpected loss at the beginning of May was a body blow to CPT, challenging even its existence. But, cometh the hour, cometh the man. Or, more accurately, cometh the woman. Helen, as we knew she would, took the helm firmly and steered the charity through the toughest times.

So when Helen asked me if I could help out a little bit at the Rallying meeting, it was an offer I could not refuse - an opportunity to talk science and catch up with some of my favourite people in Parkinson's.

Yes, I am still retired. But for a few days, I'm slightly less retired.

ARIZONA MIKE

In downtown Grand Rapids Michigan, opposite a hipster coffee shop somewhere around the intersection of Monroe and Lyon you will find him. Usually alone. Sometimes in conversation with other street people. Mostly just whittling wood all day long, fashioning canes from small branches.

Mike, or "Arizona Mike On the Block" was his name. To be honest we never expected to hear his actual name. Or his actual address. In the shadowy world of the street people identity and location matter little. Your identity is whoever you want to be and your address is, well, everywhere and nowhere. But for today he was Arizona Mike from the corner of Monroe and Lyon.

We talked a while. Mike's story was no different from that of the other street corner dwellers. A normal enough upbringing, service in the Armed Forces, a wife. A Navy veteran, he had served on the USS Enterprise. His emaciated chest almost puffed out with pride at the reminiscence. Then pancreatitis, problems with alcohol and a downward spiral that cost him his marriage and home. Not one big thing, just a sad sequence of little setbacks. The kind of things that happen daily to all of us.

Yet he held no bitterness. We talked about many things. He was happy to talk. We railed against our governments and their treatment of the disabled. Gloria took him a coffee and we sat outside the hipster coffee shop, sipping espresso and watching him whittle away at what was now to be my cane. He brought it over to us when finished, smooth and polished with a rubber toe. An elegant piece of black cherry wood. He handed it over with pride. And rightly so.

We finished our coffee, bade him goodbye and were on our way.

Mike is just another reminder of the many incongruities of this giant, sprawling country and the blind eye it turns to those on the ragged periphery of society. Trump may talk of making America great again but he will never achieve that while the Arizona Mikes of the country occupy the street corners and prick the nation's conscience.

2018

LET NO DAY BE WASTED

I don't know about you but most of my New Year resolutions are broken fairly swiftly. In the case of those involving chocolate, fairly instantaneously. My children have even taken to hiding the Bendicks bittermints from me, fearful of their potentially cataclysmic effect on my blood sugar levels. It goes without saying that their resourcefulness in sequestering said confectionery is only matched by my determination to find the little blighters. Each year they underestimate the lengths to which I'm prepared to go.

This kind of willpower and determination could, of course, be of genuine value if turned to more elevated causes. I could be out on the street corner rattling tins for the charities of my choice. Or fashioning bookshelves from MDF. Or any of the thousand and one more useful or ennobling expenditures of my time than devising a search and rescue plan for after-dinner mints. Albeit very good after-dinner mints.

When I retired last year from active service at the Cure Parkinson's Trust, it was for a number of reasons. Most ostensibly, I wanted to spend more time with my children (whether they liked it or not), having adventures and so forth. But, of nearly equal weight, I wanted to use my time in a range of more creative pursuits, writing in particular. This has been a partial success. My children bought me an Art Pass for Fathers Day which gains me free access to a great many art galleries and museums throughout Britain. Most notably, they elected to buy me a +1 version so that I could take a guest each time. I think they draw straws to see who will be my +1. It's the thought that counts.

The writing has been rather more challenging. Challenging in the sense of has-not-even-written-a-single-word. I have had the most profound sense of apathy and of writers block for several months. Although I have opinions on many things (I see your eyes roll upwards), I no longer feel the compelling urge to convey those to paper. I am happy to keep my counsel rather than puff out my chest and go looking for a fight. This is unlike me. Normally the drive to convey and convince is strong. If I had an opinion, it was important not only that you knew it but were also persuaded by it. And if that meant cracking a few eggs along the way, so be it. But it doesn't work out like that anymore. I am happy to hold my tongue rather than pick fights or batter my gentle readers into philosophical submission.

To be honest, I am at a loss to explain the sudden passivity but I do put it down to retirement. A neighbour (retired) told me that he has never been so busy and wonders how he fitted everything in before retiring. Certainly there is a sense that activities expand to fill the time allotted. It took me nearly three hours to change a plug the other day although I pin much of the time expended on my Parkinson's especially the loss of dexterity. Still three hours is a long time. And this runs counter to my ethos. I have never liked to waste time. It's a commodity in limited supply and your own supply of it is only apparent at the very end. It's easy to worry about this if you are the worrying type (as I am). I think of it as a sort of temporal anxiety.

The irony is that it is possible, by being concerned to fill one's time usefully, simply to fill it with temporal anxiety. And although one has indubitably filled one's day with thinking, it's not of the productive type. There is a saying to the effect that one should live every day as though it were one's last. I'm not sure I entirely go along with that but I do believe that one should find a way of using time effectively. But how?

So compelling is this conundrum that it has even been suggested that I come out of retirement, essentially reversing last year's decision. Although I can see the rationale, I don't believe that's a solution, for a number of reasons. But it does draw attention to one of the clear differentiations between work and retirement - the need to be useful. At home, in the bubble of retirement, one can be as useless as one wishes. The same philosophy, applied at work, swiftly curtails your employment prospects. And maybe that's the issue – I need to feel useful.

With the passage of time, I am always aware of its useful and useless expenditure and the need to differentiate each. Daytime TV is obviously useless. Sharing an art exhibition with my children is the opposite. And in between, there is a whole range of useful to useless temporal expenditure. This leads me on to my main resolution for this year. It has nothing to do with chocolate, you will be relieved to hear. In any case that would be doomed to defeat. No, my resolution this year, above all others, is to make sure that I do something useful each day whether that be erecting bookshelves, writing critical essays on Wagner or helping out as a sounding board for my many Parkinson's friends.

Don't get me wrong – I am still retired and (with both Parkinson's and diabetes) need to take things easier but I do intend to make sure that I stick to this resolution. And what is it? It's very simple really – let no day be wasted. In other words, I should be able to say, at the end of each day, that I have at least done something useful, however small. That's it. LET NO DAY BE WASTED.

THE JOY OF SCIENCE

I was not always a person with Parkinson's. For more than two decades, I was a scientist. Better than that, I was a neuroscientist. And better still, a research neuroscientist. And to be a research neuroscientist in the 1980s and 90s was to win first prize in the lottery of life.

I often reflect on my chosen area of research, wondering, in essence, whether I chose it or it chose me. With hindsight I wonder if it chose me in some way to prepare me for the second half of my life, life after research. Why? Because my main chosen area of investigation was Parkinson's, specifically dopamine function in the basal ganglia. I came to the London Hospital Medical College as it then was, in 1980 to do postgraduate research on the factors controlling dopamine function. Although I didn't know it at the time, this medical school was where James Parkinson himself had trained to be a physician. Doing my postgraduate research there had a certain symmetry.

For those of you unfamiliar with the process, doctoral (postgraduate) research is conducted under the supervision of a senior academic. And such is the intensity of the investigational process that the relationship between supervisor and student is critical to the success of the research. It is like the relationship between the quarterback and the coach on an American football team. A bad relationship renders the process unworkable. But with a good working relationship, everything else falls into place. I was blessed – my supervisor, Zyg K, was inspirational. He firmly believed that nothing was impossible. Sure, there were things that had never been done before, but that didn't make them impossible. It made them exciting. His attitude was infectious. It is a measure of the man and his influence upon me that we remain friends to this day. He set me off on a journey I would never forget.

I can honestly say that I have never been happier than when I was involved in research. There was a sense that every single new day might bring some new learning or understanding. Every sunrise was gilded with optimism, every sunset with satisfaction. To be a research neuroscientist in the 80s and 90s was to be given the key to a chamber of secrets. This was a time of huge change in neuroscience with, it seemed, new breakthroughs every other week. We were fortunate to be using techniques that were, albeit briefly, in the vanguard.

I won't bore you with the details but the gist of it was simple. Neurotransmission, the process by which nerve cells communicate with each other, takes place on a timescale shorter than a second, in milliseconds and tens of milliseconds. We had a method that could look at the changing neurotransmitters, especially dopamine, over that timeframe, in essence listening to nerve cells chattering to each other. For a few short years, everything we looked at turned to gold. Everything seemed to provide new insights into the fundamentals of neurotransmission. We published papers, spoke at conferences, spread the word in every sense.

And when the chance arose, I took over my own lab and had students of my own. We looked ever deeper at the mysteries of dopamine and its roles in the basal ganglia and limbic system. For more than a decade, we continued where my PhD had finished. I tried to instil the lessons Zyg had taught me into my own students. Several have gone on to great things, are professors and – who knows – maybe pass on some of my thoughts to their students. God help them!

I've often been asked what it is about research that made me so passionate. It's not simply knowledge. That is easily acquired with time. I think it's a number of things. Partly it's creativity. Anyone who subscribes to the notion that artists are creative and that scientists are not has clearly never been anywhere near a research laboratory. Some of the most creative and imaginative minds I know work not with paints and easels but with technology at the very limit of its capabilities. Good scientists are always asking "what if...". The best are finding ways of answering the question.

It is the nature of research that it answers questions. And it's competitive. Other labs are looking for the same answers. That lends a frisson of excitement to everything in research. Few things were more satisfying than reaching the end of a week and knowing that you had found the answer to a critical question. Over that weekend, and before the results were published, you knew that you and you alone were the only person in the world who knew that particular answer. It made you tingle.

But for me, the joy in answering research questions did not lie in finding the answer. It lay in knowing that it was the answer, that there could be no doubt.

If fate hadn't dictated otherwise, I'm sure I would still be in research. It was where my passion lay. It was where I felt alive. And if nothing else my life with dopamine prepared me for my life without.

MESSING UP THE BUCKET LIST

Every so often I revisit my bucket list. I remove activities that seem less interesting and re-prioritise the remainder. Often I just simply delete existing activities. If I haven't got round to doing them in the first sixty years of my life, they can't really be things that I want to do, can they? If hang gliding was an overwhelming priority, I would already have done it. And when you reach the point that activities are just simply being added for the sake of it, why bother?

But an entirely different matter came to a head last week. I have been meaning to go to Iceland for some time to tick it off my bucket list. I have even, more than once, made tentative plans with friends which have somehow fallen through. But last week it actually happened. I went to Iceland. We landed in Reykjavík on Monday and spent a hectic four days doing most of the obvious and immediate tourist destinations and activities.

On the first day, we ate Icelandic delicacies such as fin whale (which I promise I will never eat again and shouldn't have eaten in the first place) and – I'm not making this up –

dung-smoked guillemot (which I also promise never to eat again but for entirely different reasons). Tick.

In the evening, we were driven out to the middle of nowhere and saw the Northern lights. If I'm honest, there were a mite underwhelming but our guide assured us that they were normally much more spectacular. Still, tick.

The following three days were a melee of waterfalls, geysers, geothermal springs, volcanoes and glaciers. Tick, tick, tick, tick, tick, tick. We passed on the opportunity to rub silica mud on ourselves at the Blue Lagoon. I'm from Yorkshire and can never really get my head around this Scandinavian penchant for public nudity. The world has enough trouble with me fully clothed. It's certainly not ready for me in my birthday suit. We parked ourselves in the bar while the rest of the coach party cavorted in the lagoon. Two small beers and a bar of chocolate. £37. Yes, Iceland is unbelievably expensive. Tick.

In the four days that we were there, we barely scratched the surface of the country. Yes we saw volcanoes and so on. Yes we went to lots of places. But no, we don't feel we really "did" the country. This is not a country to be examined in twenty minute blocks, getting on and off large tourist buses. This is a country to savour at your own pace. It's a country which repays an investment of time. You need to get off the beaten track. You need to avoid the tourists. You need to get away from civilisation. And when you do, the country will give up its secrets.

I thought I would tick Iceland off my bucket list. Boy was I wrong.

A MATTER OF LIFE AND DEATH

I'm reading two books at the moment. The first is Being Mortal by Atul Gawande. The second is When Breath Becomes Air by Paul Kalanithi. Both, in different ways, tackle the subject of death and try to contextualise the experience. Gawande discusses how attitudes to life and death differ according to culture and how, in the West, we have lost sight of death as part of life. Kalanithi's book is an account of his parallel change from neurosurgeon in training to terminally ill cancer patient, examining what matters in those conflicting circumstances. Considering our collective squeamishness over the subject of death, both have sold extraordinarily well. Their presence on my bedside table is testament alone to the strength of the writing.

I have thought a lot about death recently. Sometimes in a matter of fact sort of way – do I want to be buried or cremated? Sometimes in a more spiritual way – reincarnation, belief, religion, that sort of thing. Sometimes in a personal way by the loss of a friend. In Parkinson's, we seem to lose our friends more often than we would hope. This week the Parkinson's community lost one of its strongest campaigners and I lost a friend. Not the first. And certainly won't be the last. Tina Walker was a strong campaigner for young onset Parkinson's disease. For three years we sat together on the editorial board of The Parkinson, the house magazine of Parkinson's UK. We shared a laugh on many an occasion. She loved rattling people's cages. Pretty much everyone at those meetings turned up in business clothes. Tina used to wear a Motorhead T-shirt. You couldn't help but love her. Tina was diagnosed at forty four. She was fifty nine when she died.

One of the first things you are told when you are diagnosed with Parkinson's is that it will not kill you. People die *with* Parkinson's but not *of* it. And, as new patients, we cling to this notion. We even pass it on to others as though, in doing so, we strengthen its credibility. But why do we believe this? Is it even true? Is the reality so unpalatable that we must replace it with lies? And how does this help?

Let me give you an example. People with Parkinson's often have impaired balance and are much more prone to falls. So, imagine for a second that Mr Bloggs, a seventy year old man with Parkinson's, has a fall. An ambulance is called and he is taken to hospital with a suspected broken hip. Being bedridden, it is not long before he contracts a chest infection. Antibiotics fail to control this and he dies. The death certificate lists pneumonia as the cause of death. At face value, Mr Bloggs has died *with* not *of* Parkinson's.

And herein lies the crux of the matter. Yes the pneumonia was the final straw that took Mr Bloggs away to meet his maker. But it is delusional to believe that Parkinson's played no causal role. If Mr Bloggs hadn't had Parkinson's, chances are he would not have fallen. So the Parkinson's is, albeit one step removed, very definitely causal in his death. Mr Bloggs has died as much of Parkinson's as with it. Let's be clear on that.

The simple fact is that death certificates tell only a partial truth. The death certificate is little more than a summary of finality rather than a discourse on causality. Many a complex clinical picture is hidden by the brevity of wording on the death certificate.

If you look at all-cause mortality (that's death by whatever it says on the death certificate) in people with Parkinson's, the picture is much clearer. On average, if you have Parkinson's, you will die younger than if you did not. And this is particularly pronounced for those with young onset Parkinson's (YOPD). If you are diagnosed with Parkinson's at the age of ninety, you will probably lose about a year of your life expectancy from that point. On the other hand, if you are diagnosed with YOPD at forty, you are going to lose more than a decade of the life expectancy you could have anticipated if you had not developed Parkinson's. You will live, on average, to your early sixties instead of your late seventies. I was diagnosed in my late forties. My anticipated life expectancy is a further twenty years from there, on average. I will miss out on another eight years. That's the period when I might be enjoying retirement or celebrating grandchildren. This is the stark reality of the condition. It's time we dispelled this notion that Parkinson's is not a killer. It is, and until we kill it, it will continue to be.

But life is more than numbers. Mozart was thirty five when he died. But in that short life (even by the standards of the eighteenth century) he wrote some of the greatest music ever performed. He filled his life to the brim. Which leads me back to Atul Gawande and Paul Kalanithi. Life is not about length but about value and quality. Kalanithi especially shows that life is even more important, vital in its truest sense, in the face of death. We don't judge cricketers by how long they occupy the crease but by how many runs they score or wickets they take. It falls to us, in our shortened lives, to make them extraordinary – to be the best we can, to suck the marrow out of experience. And we need to look reality in the eye.

Carpe Diem.

Further Reading

Atul Gawande (2014) Being Mortal

Paul Kalanithi (2016) When Breath Becomes Air.

Lianna S Ishihara, Anne Cheesbrough, Carol Brayne, and Anette Schrag (2009) Estimated life expectancy of Parkinson's patients compared with the UK population. J Neurol Neurosurg Psychiatry. 78: 1304–1309.

WINNER TAKES ALL

At first sight, the pharmaceutical industry seems readily understandable. There is a need for drugs to treat illness. Drug companies collectively provide these treatments. And as time progresses they produce better treatments. This is big business. Take Parkinson's: the estimated current annual cost of drugs to treat the condition is $4.24 billion in the US alone [1]. Whilst not as lucrative as fields such as cancer, it is nonetheless a reasonable contributor to the balance sheet of the average pharmaceutical company. Couple this with the predicted rise in prevalence as the population ages, and this amounts to a reliable and predictable income stream for the foreseeable future ($5.69 billion in 2022 [1]).

Pharmaceutical companies answer to their shareholders. They are not philanthropic organisations. Without profit, they don't exist. But as long as they produce new and/or better treatments, they continue. This is simple business and we, as patients, should not begrudge them reasonable profits. The alternative is the cancellation of research programs and ultimately a reduction in choice for patients. And none of us want that.

The drugs we have currently available for Parkinson's have one thing in common – they are all symptomatic treatments. They ameliorate symptoms, or attenuate the side-effects of other drugs used to treat Parkinson's, but do nothing (with the faintly possible exception of rasagiline) to slow the progression of the condition.

Not surprisingly there are, particularly among the patient community, those who feel that a cure for Parkinson's will not arise - for the simple reason that the pharmaceutical industry has no interest in developing such drugs or approaches. Indeed, many feel that there is a positive disincentive for drug companies to develop a cure. And it's easy to see how this thinking evolves.

A cure for Parkinson's, or even a drug that halted progression, would have significant implications for sales of symptomatic treatments. Development of a cure by the pharmaceutical industry would, by this logic, be tantamount to biting the hand that feeds. On that basis, drug companies might be expected to close ranks and implement a policy, implicit or explicit, not to develop disease modifying agents.

Now I'm not saying for one second that such a policy exists. On the whole, I don't subscribe to conspiracy theories. But the current situation is, it has to be said, convenient for shareholders in the pharmaceutical industry. There is a steady supply of patients and the drug companies can continue to meet that with an equally steady supply of new symptomatic treatments. A cure for Parkinson's would, in financial terms, be decidedly inconvenient. Although literally life-saving for patients, it could amount to a serious financial knock to those sectors of the industry making only symptomatic treatments. It's easy to see that companies would be reluctant to invest money in avenues that would potentially cut off their own income streams.

However this position is untenable, being based on the false premise that any cure, were it to come, will emanate from the pharmaceutical industry. For the most part, and bearing in mind the amount of time it takes to develop a new treatment from lab to patient, this has historically been true. Only drug companies have had the resources to conduct such work.

But the old models of drug development through cellular approaches, screens in laboratory animals then healthy volunteers and ultimately patients no longer hold. Drugs are no longer developed through such rigid methods. Increasingly the regulatory agencies are open to other approaches, particularly the repurposing of medications. And the patient voice is growing louder and louder. Patient want cures. The doors are open for academic institutions and the charity sector to engage in drug discovery. And to do this to an agenda driven by patients.

This changes everything. The pharmaceutical industry no longer holds a monopoly on drug discovery. The introduction of other players into the arena changes the rules. And the rules no longer allow for 'olde worlde' symptomatic drug treatments. The race is on for a cure. And that cure, whisper it, may not come from pharma.

Think of it as a cake. Instead of a modest Victoria sponge being sliced up among many companies, we have a single fabulous cake, a gigantic Sachertorte, as the prize. And that prize will go to one destination - the institution, charity, university or drug company that develops a cure for Parkinson's. The pharmaceutical industry has a stark choice – either join the race or lose by default. After all, shareholders are tough taskmasters but they understand the rules. And the new rules are very simple. Winner takes all.

[1] https://www.marketsandmarkets.com/PressReleases/parkinson-disease-treatment.asp

SPEAKING IN TONGUES

There are times when I like to delude myself that I am a serious writer, that I can string words together in ways that touch readers. I like to feel that my words have impact, that they demand reaction. This is of course a Walter Mitty conceit on my part. Every once in a while I write something which seems to evoke a visceral response amongst my readers. And I cherish those moments. But they are infrequent. I was going to say that they are unpredictable but that is not the case. There are certain subjects I can touch upon in the full knowledge that I'm pressing buttons I know to trigger responses in everyone, irrespective of perspective. Speaking in tongues, if you will. And that in turn makes me reluctant to press those buttons. It seems cynical. At best, it is a revisiting of familiar pastures. And I have always been of the opinion that one should take the path less travelled.

Mike Oldfield had a huge success with Tubular Bells but progressively less with each subsequent album. It's not that the music was necessarily weaker, just the presentation. Eventually he hit upon the wizard wheeze of calling each new album Tubular Bells part two, three and so on and making brief musical nods to the original in each. Cynical maybe, desperate certainly.

There are millions of opinions on what makes good writing. When I wrote my PhD thesis in 1984, I was advised to "write like Hemingway". There should be no wasted words. Needless to say, only Hemingway could write like Hemingway. And to assume another person's voice is, even if done well, uncomfortable. I tried nonetheless but still managed to exceed the prescribed fifty thousand word limit by a further twenty six thousand words. Thank God my examiners never took it upon themselves to count. Although my early scientific papers were lean and mean, I gradually found an easier and, for me, more authentic style.

I wrote a lot at school, occasionally picking up prizes for English and so forth along the way. But apart from scientific documents, I had little cause to do so thereafter. I'm not sure I even felt the urge. Until of course I was diagnosed with Parkinson's. That event, which seemed to marry up science and personal experience, lit a creative fire under me. Suddenly I enjoyed writing again. Readers wrote to me and told me they enjoyed reading my pieces. Sometimes I touched hearts and sometimes I touched nerves. But I still never considered myself to be a writer. And I still don't.

To be honest, I read little these days for a variety of reasons. I struggle to read the work of an author I admire without my own writing subtly absorbing their mannerisms. It's not intentional or contrived in any way, which somehow makes it worse. So at various stages, my writing has been imbued with tiny fragments of Hemingway, Kerouac, Proulx, Keillor, Bryson, Steinbeck and Faulkner. Interestingly all

Americans I notice as I write this. I don't know whether these have been aids or barriers to finding my own voice.

Funnily enough, the route to finding my own voice has been to use my voice. Before the advent of speech recognition software, I wrote (painfully slowly) using a keyboard. By the time I had finished a sentence, I wanted to change it. My brain outran my fingers, even more so as Parkinson's tightened its malign group on my digits. Now I use voice recognition software for everything. And in the truest sense, I feel I have found my voice. I think, I talk, it writes. And my natural voice as a writer emerges, butterfly-like, from this confining chrysalis. Sure it's wordy and rambling. I drift off down blind avenues. But it is me. I may not be a writer's writer. Probably never will be. But this is my voice.

I have written two pieces in my life of which I am genuinely proud. The first, a short story called Angel is on my website. The second was a five hundred word catharsis written the night my mother died called The Hardest Goodbye (not currently public). On those two pieces, I considered myself "a writer".

That's a total of two pieces in a decade. Less even than the frugal Salinger. I have good friends who are excellent writers. And occasionally I can come, like Cutty Sark, within grasping range of their abilities. But for the most part, I still consider myself an aspiring writer.

At sixty, I'm probably leaving it a bit late.

THE CANDLE AND THE FIRE

It is a year since the Parkinson's community lost one of its strongest advocates, a full year since Tom Isaacs passed away unexpectedly early one May morning.

I remember the phone call from Helen, mid-morning. She said hello and asked how I was. I recall thinking that her voice was very flat. There was a pause before she told me she had some devastating news. Tom had died that morning. I felt numb instantly. I don't remember what I said but I knew that I needed to get off the phone immediately. That maybe, if I put the phone down, I could pretend it hadn't happened. After all, I had been speaking to Tom only a couple of days previously. How could he be dead?

It's all a blur now but, as the news sank in, I remember wanting to go up to the CPT office in London simply to be around Helen, Anna and the others – my people – and to have them around me. We all had our stories about Tom.

It was not long before word seeped out into the Parkinson's community. Not long before people started phoning the CPT team and myself expressing their condolences and asking questions about the charity's future plans.

And over the following weeks and months, something rather remarkable happened. Visits to the CPT website rose sharply and donations to the charity increased. Somehow Tom's death had pricked the public consciousness over this still incurable disease. And, even more remarkably, callers were looking for ways to turn their donations into something positive and enduring. Many spoke of Tom's legacy of patient advocacy and of his role in the genesis of Parkinson's Movement. And of Tom's unwavering belief that research, and research alone, would eventually result in a cure for Parkinson's. There was a tidal wave of support for CPT and a general sense that, even if Tom was gone, his work would not only continue but at a greater pace. This is Tom's legacy.

A lot of people ask me what he was like to work with. Exuberant, exhilarating, exasperating. All of those things in one. Always challenging existing dogma. Always asking "why not?" questions. A man for whom "no" was never an acceptable response. Tom always liked to push your buttons. He and I would spar regularly on almost everything, running ideas past the other. We never gave each other less than complete honesty. He was never less than inspirational. It was a privilege to work with him. I miss that. But more than that, I miss those moments, towards the end of the day when we would talk about everything except Parkinson's. We shared some of our other thoughts. So yes, it was a privilege to work with him but it was a greater pleasure to call him my friend.

And his legacy continues to grow. To quote a line from a Peter Gabriel song about another activist, "you can blow out a candle but you can't blow out a fire". Tom may have been the candle but CPT and PM were the fire he started.

HAVING MY CAKE AND EATING IT

Fourteen months ago I made the decision to "retire" from the Cure Parkinson's Trust, where I had served for five years. It was a tough decision to say the least. CPT has been my "other" family for nearly a decade, sharing my trials and tribulations as well as my successes. But I felt I needed time to myself, time to share with my children and time to write and enjoy my other pastimes. I think also, in a way, I saw it as a kind of retirement from Parkinson's. At least symbolically.

I am someone who likes to have their cake and eat it. I like to be 'in the loop' without necessarily being weighed upon. I like to be wanted but not needed. I like to have all the perks of involvement without the burdens. Does that sound greedy? I suppose it does. And in that respect, I suppose I felt that I could achieve a measure of detachment whilst simultaneously still being, at least intellectually, involved.

But retirement doesn't work like that. As time passed, I found myself increasingly peripheral. Others took up my roles with aplomb. As the months passed, so apparently did

I. Up until my retirement, every day brought invitations to speak, requests for opinions, offers of involvement. As the months passed, I gradually realised that retirement carried a cost. Fewer people wanted my opinion. I had more time to reflect and more time to opine. My thoughts were more clearly formed. Yet, in that most sublime irony, nobody was interested. Although, in some ways, I had my cake, clearly others were eating it.

Retirement meant intellectual redundancy. And I didn't like it.

Of course the notion of retirement from Parkinson's is intrinsically flawed. If I had seriously entertained the notion that I could retire from Parkinson's, I was mistaken. Parkinson's, after all, showed no inclination to retire from me. It remains its usual unwelcome self, always reminding myself of its chameleon like presence. One day it's dystonia, the next dyskinesias. Pain one day, stumbling and reeling the next. Never short of creative ideas to make my days just that little bit less pleasurable than they need to be.

So what is the solution?

Do I want to stay retired? Yes. At least a qualified 'yes'.

Do I want to still be involved? Ditto.

I want everything. I want to be wanted still. I want to be heard. I want my thoughts to be needed. I want people to listen.

Yes, I want my cake. And I will jab a cake fork into the hand of anyone who tries to eat it!

AWARENESS, ADVOCACY, ACTIVISM AND ACTION

There exists a relationship between patients with any condition and the researchers. For some conditions there is barely an acknowledgement of each other. For others, such as AIDS, the relationship is much tighter, with a strong knowledge interchange between the two communities.

In Parkinson's, as in so many other conditions, we use terms such as awareness, advocacy and activism often rather broadly. Sometimes even synonymously. Yet, to my mind, there are subtle but important distinctions between the terms. Awareness of the condition is essentially the entry point into the whole spectrum of patient engagement. It is the point at which the person with Parkinson's begins to develop an understanding, both personal and general, of the condition. Advocacy is a level up from there. In the context of Parkinson's, advocacy can be defined perhaps as public endorsement of the importance of research. It is the point at which the person with Parkinson's changes from being a net recipient of information and/or benefit to a net contributor, in essence sharing their knowledge and experience. Activism can, on the one hand, be seen as synonymous with advocacy but, to my mind, the term has slightly different connotations. Activism is a more dynamic term, suggesting greater vigour. One might almost think of it as a more forceful, aggressive even, form of advocacy. Activism is also perhaps the form of advocacy most likely to effect action.

Let me explain.

The AIDS community is often held up as an icon of the power of patients to exert change. This change was not achieved by gentle advocacy but by forceful activism. A simple Internet search for the term "AIDS advocacy" generates 61,900 hits, while a search for "AIDS activism" finds 186,000 pages, more than three times as many. This is of course a crude measure but nonetheless makes a clear point – people with AIDS are more activists than advocates. And this has translated into tangible results – better access to drugs, improved treatment paradigms and hugely enhanced public understanding of the condition. Indeed, the AIDS community were very much involved in translating AIDS from the death sentence it initially was to a long-term treatable condition.

Similar searches, in the context of Parkinson's, are illuminating. "Parkinson's advocacy" yields 5210 pages while "Parkinson's activism" finds a mere 42. In essence, although we have a community of advocates, we are not necessarily seen as activists. Although it is a simplistic analysis, one could suggest that this inertia is hampering research. The researchers do not know what we, as people with Parkinson's, see to be the research priorities. Following the AIDS model through, more activists translates into greater action. We are not making the transition. To my mind we need to take that step from polite advocacy to more forceful activism. Because that's how we will exert action.

The AIDS community has shown us the way or, at least, a way. We need to learn from their successes and failures how best to exert leverage on Parkinson's research and researchers. How can we most effectively make our issues their issues?

We need to come up with the answers.

WHY HAVE WE LARGELY IGNORED NON-MOTOR SYMPTOMS?

I often have the feeling, rightly or wrongly, that medicine is happy to investigate those areas where it is more easily successful than those which prove more challenging. These are in essence the low hanging fruit, easy wins for impatient researchers. And in a research culture where early success or failure can build or blight a career, it's easy to see why some areas of research are more fashionable than others.

In the context of Parkinson's for instance, we have known about rigidity, bradykinesia, postural instability and tremor. We have known about them because they are easy to identify, easy to characterise and, in simple clinical terms, easy to treat. Results are visible and readily quantifiable by physicians. They are, for Parkinson's, the low hanging fruit.

Further up the tree, and less readily accessible, are some of the non-motor symptoms of Parkinson's. These might include fatigue, affective disorders, bowel and bladder control issues, apathy, sleep disorders, loss of sense of smell and so on. And although often nebulous and, in many respects, non-specific, they nonetheless form key parts of the Parkinson's experience. Indeed for many patients nonmotor symptoms have a larger role in determining quality of life than the motor symptoms.

Yet even nowadays nonmotor symptoms tend to be thought of as an adjunct to the central motor symptoms of Parkinson's. Somehow they are not considered still to be central to the condition. Even now in 2018 we persist in calling Parkinson's a movement disorder rather than the broad-spectrum neuropsychopathological condition that it is. We seem to insist on the pre-eminence of motor symptoms.

But why should this be? Why should we persist with this archaic terminology and its stultifying implications for our understanding of Parkinson's?

A number of possible explanations come to mind but I like this one – that by placing undue emphasis on motor symptoms, the condition is medicalised. That is that it is taken out of the ownership of the patients and becomes the province of the treating physician. When it is a motor condition, a movement disorder, it is easy for a treating physician to claim therapeutic success and easy for a researcher to study. A reduction in the amplitude of tremor is something numerical and therefore scientific. For nearly 2 centuries since its original characterisation, Parkinson's has been a movement disorder that is managed by physicians with scarcely a nod towards the nonmotor symptoms.

There is a tendency to believe that the nonmotor symptoms of Parkinson's are something new or at least newly identified. The derivation and application of scales with which to quantify the nonmotor symptoms has taken them into the spotlight, the point at which the medical research community has begun to show interest.

But nonmotor symptoms are nothing new. In his original essay on the shaking palsy, Parkinson described nearly all of the currently understood nonmotor symptoms of the condition. Parkinson knew they were part of the condition, because he asked the right questions of the patients. Since then nearly 2 centuries worth of physicians have found it more convenient not to ask the right questions. So although nonmotor symptoms have been part – a central part – of the patient experience of Parkinson's, they have only more recently become legitimate targets for researchers and physicians. Before they were merely irritating distracters, blemishes on their motorcentric views of the condition.

Nonmotor symptoms are often difficult to treat. It's relatively easy to improve mobility or reduce tremor but apathy and fatigue remain less amenable targets. But if you ask any patient, you will find that the nonmotor symptoms impinge every bit as assertively on a patient's quality of life as their mobility. They may be more tricky to treat. They may be less readily quantifiable. They may even be elusive research targets. But they are vital components in the Parkinson's jigsaw.

I was a researcher in Parkinson's for more than 20 years before I was a patient. More importantly, I had to become a patient before I really understood this condition. My dozens of research papers as a scientist were testaments to what we could measure in Parkinson's, not what we should.

We have gorged for too long on the low hanging fruit. It's time to set our aspirations higher.

COOKY AND JIMMY

Anyone who believes that test match cricket is a dull, dreary, ritualistic relic from another time should have witnessed the events at the Oval today. This was test match cricket at its finest, full of intensity, passion and commitment. This was drama on an epic scale. Drama that bore no relation to the artificiality of limited over cricket with its contrived tensions.

I have little interest in limited overs cricket and none whatsoever in 20/20. It is a cruel bastardisation of the true game forced upon the viewing public by television grandees anxious to fill football match size slots in the evening schedules. Even the language we use to describe it is insidious. We talk about the longer and shorter forms of the game in the same breath, as though each were an equally valid representation of the game.

I question this. Has our appreciation of the infinite nuances and subtleties of 'the longer game' dwindled so far that we would happily fork out our hard earned shekels to watch 120 ball nocturnal slogfests played, appropriately enough, by men in pyjamas. This is postcard cricket. A tiny little vignette, artificially coloured, purporting to represent the true game.

But where is the intensity? Where is the drama of watching the greats of the game battle it out over four or five days? Where is the ebb and flow, as wickets fall and runs flow? Where is the appreciation of quality strokeplay? Cricket is a game for reflection and observation, not cheerleaders, dancers, fireworks and endless repetition of Queen's "We are the champions". This is not a celebration of cricket. This is its death knell.

Richard Benaud, the doyen of commentators, articulated the differences clearly. Limited overs cricket, he said, is an exhibition. Test match cricket is an examination.

Today we saw test cricket at its finest. We saw a match between England and the top-ranked test cricket nation, India, swinging to and fro as ascendancy moved from one country to the other. Any result was possible the beginning of the day and remained so until perhaps the last couple of hours. A titanic contest between England's bowlers and India's batsmen drawing the drama of this excellent series to a conclusion.

But more than this, today was a day when international cricket said goodbye to Alastair Cook, England's rock of an opening batsmen, former captain and most capped player, the highest scoring left-hander in test match history and scorer of over 12,000 runs. The standing ovation he received on reaching his last test century probably left a lump in many throats. You couldn't have written it any better.

But the scriptwriter wasn't finished. As the last hour of play began, Jimmy Anderson needed one more wicket to overhaul Glenn McGrath's world record of 563 test wickets by a seam bowler. Over after over passed as Jimmy beat the edge to no avail. Rashid and Curran took wickets until there was only one left to take. Eventually Root gave him the new ball and Jimmy made history, uprooting Shami's middle stump.

There were tears, speeches, champagne and smiles. This was drama, played out by men in white – the culmination of the summer's battles between two great teams. It was cricket at its best, a team game played by individuals. This was the drama of characters not caricatures. The next time we feel inclined to watch the men in pyjamas, perhaps we should remember what we saw here today. This is test cricket and it's called test for a reason.

SELF EXPERIMENTATION IN MEDICINE –
MY CHOICE, MY DECISION

Anyone who has ever tried to conduct medical research will be familiar with the workings of the ethical committee, an institution appointed arbiter of what research is or is not acceptable. Pretty much all institutions that conduct medical research have an ethics board. For the most part this makes sense – it is there to protect subjects in clinical trials from unnecessary dangers, whether physical or psychological.

On the whole, ethics committees are effective and give researchers the opportunity to reflect on their work and, in many cases, to improve upon the experimental design. The committee seeks to protect individuals from ill considered experimental protocols. It is the ethics committee that decides whether a given study can get by with a single lumbar puncture rather than the half-dozen proposed by the investigators. This is an extreme example but you get my drift. In essence, their role is protective.

In extremis, the ethics committee has the power to block the research from taking place at all where there are seen to be avoidable risks to the patient or where the scientific benefits do not outweigh potential dangers to the patient. On the whole, ethics committees are on the side of the patient rather than the experimenter.

But what if the patient IS the experimenter? What if a patient wishes to conduct research using themselves as the subject? What does the ethics committee do under these circumstances? Does it attempt to protect the patient from themselves, vetoing such investigations? And if so, under what authority does it presume to act?

The medical literature is full of examples of scientist physicians experimenting upon themselves. Take for instance Sir Henry Head FRS, whose pioneering work on sensory nerve reinnervation was based largely on the results of severing the radial and external cutaneous nerves in his own arm. Yet no ethics committee on earth would allow him to conduct such experimentation on others. So Head simply did the work himself and, in doing so, significantly advanced the field of research.

In the case of Werner Forssmann, he sought ethics committee approval for his work first and was refused. Undeterred, he proceeded anyway with his work on cardiac catheterisation, being dismissed from his job on more than one occasion. Perseverance eventually saw him awarded the Nobel Prize for medicine in 1956.

You might feel that these occurrences are historic and, in the light of these, that modern day ethics committees recognise the right of individuals to experiment upon themselves. You would be wrong.

A friend of mine has been working in the field of self-monitoring and self tracking in Parkinson's disease for some years. Her work has been widely reported and she is a strong advocate for the role of patients in self-care for chronic illness. Not surprisingly, her work has involved an element of self experimentation and some of this work has been published in peer-reviewed journals.

She is currently writing up a doctoral thesis and includes some work on self experimentation. This is a significant piece of work and a validation of the entire spectrum of self-care. There are many in the field who look forward to the dissemination of this work and its submission as a doctoral thesis. The work will be a credit to the institution where it was conducted.

Hard to believe then but the candidates host institution has denied her the right to submit her thesis for examination based on this work. The institution has taken the view that those papers based on self experimentation are inadmissible because the candidate did not seek ethics committee approval. That's right – although these papers have been published, following peer-review, in reputable scientific/medical journals, the institution still feels that the work is ethically unacceptable.

This is extraordinarily backward thinking. A return to the old patrician days of medicine and a vote for scientific censorship of the worst kind. If this was a new institution, finding its way in the moral maze of medicine their attempted action might be comprehensible. But the Institute in question is more than 200 years old. They should know better.

It is increasingly recognised that patients are an essential part of the research process and that their perceptions of needs should be prioritised. And where patients are winning battles for control of their own data, assessment and treatment, this action is a timely reminder that some institutions seem determined to be forever dinosaurs.

HOPE IS BACK ON THE AGENDA

The average person with Parkinson's lot is not a happy one. An increasingly acknowledged litany of non-motor symptoms add to the obvious freezes, shakes, rattles and rules that make up our daily repertoire of silly walks. We shuffle and glide, mambo and tango our way for the entertainment of others it seems. Fortunately our inability to "just walk properly for goodness sake" comes with a thick skin to protect us from the slings and arrows of our outrageous misfortune (to slightly misquote the Bard).

Each day represents another minor erosion of vulnerabilities, another minor pique to add to the already bristling catalogue of biochemical insults we endure. Parkinson's plays grandmother's footsteps with us. We stare resolutely ahead while all the while Parkinson's creeps stealthily up behind us. It's a cunning little so-and-so.

But something's changed. And I can't quite put my finger on it.

Let me backtrack a little. Over the last few weeks I have found myself on the road talking to scientists and people with Parkinson's. Okay, what's new – I have done that many times before. Indeed that is pretty much what I do. I talk to people who have Parkinson's, scientists who research Parkinson's, clinicians who treat Parkinson's and the general public which, for the most part, is very little interested in Parkinson's. And from each of these constituents I form a picture of the condition, a mental map of Parkinson's and where it's going.

The last couple of months have been quite illuminating. I have been to Sweden and talked with an old friend, a gutsy and provocative long-term advocate about the role of ethics in self experimentation. I have talked to a visionary laboratory head in Switzerland determined to make sure that patients have a real and not tokenistic role in their research endeavours. I have sat in on the machinations of the Linked Clinical Trials's august program committee and occasionally chipped in. I have been to the annual Rallying to the Challenge in Parkinson's Disease conference in Grand Rapids, Michigan, its usual heady brew of patients and scientists jostling for each other's attention. I have appeared in a webinar about how Parkinson's begins and a Facebook live session about the impact of nonmotor symptoms. I've been a busy boy. I've talked and listened, spoken and heard. I have met friends old and new.

There is a new buzz. A new sense that things are happening. The scientists are beginning to piece together the different strands of the Parkinson's story – the role of alpha synuclein in neurodegeneration, where it fits in the jigsaw and how we might manipulate that. The involvement of inflammation in some of the processes and the window of opportunity that exists for us to find drugs to block the spread of neuronal damage. And there seems to be a wagon train of information supporting the hypothesis that Parkinson's doesn't start in the brain. People are looking much more closely at the gut and at the bacteria within the gut. Hard to believe perhaps, that tiny microbes in our bowels maybe where the war that is Parkinson's actually starts. For an enlightened handful, this seems logical and credible. Yet even as little as a year ago the majority were sceptical. In around a year this has gone from science fiction and conspiracy theory to mainstream thought and research.

What does this all amount to? Are we making the picture clearer or simply muddying the waters?

I can sum this up in a single word. Hope. Or perhaps I should write that as HOPE! For decades we have looked for simple solutions and reasons to cling to hope. As the late Tom Isaacs said "I suffer from two illnesses – Parkinson's and hope. Only one of those is incurable".

And you know what – he was right. Hope is incurable. And it's back on the agenda.

A RIPE OLD AGE

Where does old age begin? Don't worry – I won't be asking questions. It's sort of rhetorical. I've been wondering but without really reaching a conclusion.

My father defined old age as anything beyond forty. Until he was forty. Then it was anything beyond fifty and so on. I remember as a child being decidedly unsettled by his frequent protestations that life beyond forty wasn't worth living and that, immediately upon attaining that landmark, he planned to put a revolver to his head. I honestly believed him. Mother reassured me – "he doesn't have a gun", which although mildly comforting was not really the point.

He regarded every tiny failure as a portent of impending intellectual apocalypse. He had only to fluff one clue in the

Telegraph crossword to turn from genial parent into glum fatalistic octogenarian. Each tiny erosion of his cognition had the same effect. One minute he was reeling off Henry VIII's wives, the next he was staring gloomily into his morning tea. It didn't take much. A single momentary slip over a deceased aunt's maiden name and he was immediately loading his imaginary revolver. And he was still only in his thirties.

Of course, with Parkinson's, you get to fast forward a little. You can skip read some of Shakespeare's normal stages and go straight from moderate self contented good health to stumbling hunchback in less than the time it takes to finish an average test match. Or so it seems.

The most obvious barometer of my old age is the perception of my children. They have, almost imperceptibly, gone from talking to me to talking about me. I exaggerate to make a point but the difference is clear if mercifully infrequent still. "Are you okay?" has gradually become "is dad okay?" I don't think they even notice and they certainly don't mean it unkindly. Indeed I think the opposite is true. I think it is a measure of their affection that they confer amongst themselves how best to help me.

This has practical implications of course. They no longer allow me to climb ladders, to clear gutters or to paint underneath the eaves. They make the point, firmly but gently, that I am no longer a fully fit thirty-year-old (actually I don't think I ever was) and that the consequences of any fall from height are likely to involve broken bones and a severe outbreak of Itoldyouso. At the very least. And worst case scenarios for hospitalised Parkies are... Well, let's not go there. Suffice it to say that you can cancel the papers. So ladders are out.

So too, it transpires, are power tools. You name it and I'm not supposed to use it. Jigsaws, angle grinders, power drills, staple guns and so on. There seems to be a blanket ban. Even the leaf blower is questionable. I ask you, how much damage can anybody do to themselves with the leaf blower? At least I now know to keep stumm about the chainsaw in the tool shed.

So what is my point? Well mainly it's this - that Parkinson's somehow ramps up the perceived speedometer of infirmity. That people judge you as older than you are. I am 60, soon to be 61 (and happy to share my birthday list with you incidentally). I am weaker than I was, of course, but not yet ready to put away my power tools even if it does mean my children cowering behind the curtains waiting for the sound of arterial blood splashing on the patio, or holding their breath in anticipation of the next accidental amputation.

Of course old age is really just that little bit beyond your actual age. My father made it to 85. He never did find that gun.

WHEN JON INTERVIEWED DR STAMFORD

Jon: First of all, some introductions. My name is Jon Stamford. I am a person with Parkinson's, diagnosed 12 years ago. It's my pleasure today to interview Dr Jonathan Stamford, former head of the Neurotransmission Lab at the Royal London Hospital. I wonder if I could start by asking you, Dr Stamford, to tell us a little bit about your research.

Dr S: Certainly. My laboratory's research was directed towards the monoamine systems of the brain – that's dopamine, noradrenaline and serotonin – and the processes which govern their actions. I was particularly interested in the control of dopamine function in the basal ganglia, the area of the brain most strongly involved in Parkinson's disease. We studied this with microelectrodes implanted into brain tissue.

Jon: And presumably the results obtained from these microelectrodes gave you some good insights into Parkinson's disease?

Dr S: Yes and no. They revealed quite a lot of information about the area around the dopaminergic synapses in the striatum. We learnt a great deal about what caused dopamine release, how far the dopamine travelled beyond the synapse, how fast those processes were and how they could be modified by the drugs used to treat Parkinson's. For instance L-dopa increases the amount of dopamine release for every action potential. Other compounds have the opposite effect. It's all valuable information into how the brain behaves at its most simplified form.

Jon: You said "yes and no".

Dr S: Yes, I meant that there were reservations. As I said, these kind of experiments are very helpful in elucidating the mechanisms of neurotransmission and finding mechanistic explanations for neurological phenomena. So they can provide insight into some of the biochemical changes in Parkinson's. And that's very helpful. But we also have to recognise that a piece of brain tissue in an organ bath is not the same thing as a person with Parkinson's.

Jon: I agree. It often seems to me that scientists can lose touch with patients.

Dr S: Very much so. And I think it's to their detriment. I enjoyed my research very much and we learnt some useful and meaningful things. But that same focus on the minutiae of neurotransmission and of Parkinson's sometimes takes you away from the bigger picture. Those tiny little details in that synaptic soup of neurotransmitters ultimately doesn't tell you much about Parkinson's patients and their concerns and needs. It is perhaps the ultimate irony that the first person with Parkinson's that I met was you.

Jon: Yes, I remember. I remember being diagnosed as well. The consultant was telling me all about these things that were going to change my life and you kept interjecting with inappropriate questions about dopamine.

Dr S: They were important to me. I was trying to find points of reference that I could understand. I knew about dopamine but I didn't know much about tremors, stiffness and balance problems. I was simply trying to relate one to the other. To try and bring my world and my experience to bear on your world and your experience.

Jon: Well, with hindsight, I think you needed to listen more to me. It didn't really matter whether you could relate my experience to your dopamine. I knew what I was feeling and it was quite difficult to translate that into language that had meaning for you as a scientist. So it's not surprising that you were struggling to convey your science to me. And in a lot of ways it didn't matter. My experience was genuine – it really didn't matter whether you could relate it to dopamine or not. Things are real whether or not you can explain them.

Dr S: I think I was just trying to make sense of it all and to find ways in which my knowledge would be helpful. Ways in which it might cushion the blow of diagnosis.

Jon: ...and did it?

Dr S: Not really but it did have one valuable side-effect. It made me aware of just how separate the worlds of science and patient experience are. It made me realise that your experience as the patient is vital to me as a scientist. Let me explain.

Jon: Please do, Dr Stamford.

Dr S: I think the patient community looks to science to provide explanations and that's not always possible. As a scientist, I know that that's why we do research – to help us provide explanations. But you can only do that if you are in touch with patients. My work was very far removed from patients. It was very much microscopic detail. As I progressed, it was a case of knowing more and more about less and less. And I think that's the exact opposite of what one should be doing as a scientist. I think it's important that we bridge that gap between scientific mechanism and patient experience. If I had seen patients while I was doing my research, I'm confident that I would have taken different directions. But I didn't have that connection.

Jon: So it's up to you as a scientist to make that connection.

Dr S: No, you don't get off that lightly! You forget that we are in this together. So it's also down to you to try bridge that gap. We need to find ways in which scientists and patients can interact. Ways in which scientists can put their work in context, explain it and its value to patients. We also need to find ways in which patients can truly convey what they are experiencing in language that the scientists will understand.

Jon: I've always thought that patients and scientists speak a different language!

Dr S: Yes they do. But my point is that they don't have to. We need to bridge that gap. We need to find ways in which to find common language. And that language is often born out of common experience.

Jon: What you mean by common experience?

Dr S: In essence the context in which we do our work as scientists needs to be broadened. We need to find ways in which we can experience the wider patient experience. We need to know what it feels like to be diagnosed with the neurodegenerative illness. How that impacts your life from thereon. We need to understand what you go through.

Jon: In essence, you need to walk in our shoes?

Dr S: Yes, I believe we do. But I also believe that you need to do the same. It's not simply a case of scientists learning what it's like to be patients. It's just as important that patients try to learn what motivates scientists. If we need to walk in your shoes so to speak, it also follows that you should put on a lab coat. We, as scientists, should find ways to communicate what drives us to do what we do. I think reciprocity is the issue here. It's all about communication. It always has been.

Jon: Dr Stamford, thank you very much.

Dr S: My pleasure.

WHAT IF?

Let's put aside the politics of Brexit and all the name-calling from both sides. At the time of writing, it looks increasingly unlikely that a deal will be obtainable. At least obtainable on anything approximating to favourable terms for the UK.

The unthinkable alternative of a 'hard Brexit', where we crash out of Europe without any agreement is increasingly not only thinkable but probable. So let's assume that March 2019 comes and goes. So what, I hear you ask. Why should it matter that we keep our own goods and they keep theirs? The answer is very simple – food and medicines. Foodwise, the best we can hope for is probably less choice, higher prices and unpredictable supply. But these deprivations pale into insignificance compared with the elephant in the room. I'm talking about the supply of medicines. It's one thing telling customers that they can't buy Parma ham or Roquefort, quite another explaining that they cannot have their drugs.

I have a chronic illness, well two actually, which mean that I have to take drugs for the rest of my life. I take eight different medications seven times a day in different combinations. Fourteen tablets, capsules or patches a day. Every day. That's 5110 doses per year. And I do not take these in order to feel a bit better or make my condition(s) more comfortable. I take them to stay alive. Without these medications, I would be unable to walk, talk, eat or take care of myself. Without these medications, my life expectancy would be a matter of months. And pretty unpleasant months at that.

I'm not alone in needing my medications. The UK has around 150,000 people with Parkinson's. But look further - think of insulin-dependent diabetics, people with epilepsy, those with chronic pain or undergoing chemotherapy. There are more than 300,000 type 1 (insulin-dependent) diabetics in the UK. 600,000 have epilepsy. These are not people whose quality of life is improved by medicine. These are people whose lives depend on medicines. It is not a "nice to have" but an absolute prerequisite. Without drugs, lives are in danger.

Do I have cause for concern? Yes, I think I do. The UK, you may be surprised to hear, does not produce much in the way of medicines. We are a net importer. From Europe in particular. Around 90% of the drugs used in the UK are imported, 45% from the EU [1]. UK imports of pharmaceutical products from the EU were worth around 26 billion US dollars in 2017 [2]. So if you think a hard Brexit is bad news for supermarkets, imagine what it's going to do for healthcare.

The NHS is, I understand, making efforts to stockpile key medicines at the moment. That is at least a start. Let's hope that we can at least buy ourselves time. Because we won't be able to buy anything else. I truly hope I am wrong. I want to believe that this can still be resolved. I want to believe that post Brexit Britain is at least somewhere safe to live. I don't expect a cornucopia of new foods and better drugs. But I do hope at least for survival.

[1]   Brexit could hinder medicines supply to UK, MPs told   https://www.pharmaceutical-journal.com/news-and-analysis/news/brexit-could-hinder-medicines-supply-to-uk-mps-told/20204082.article?firstPass=false

[2]   Value of pharmaceutical products imported into the United Kingdom (UK) from the 27 nations of the European Union (EU-27) from 2012 to 2017 (in million U.S. dollars) https://www.statista.com/statistics/497337/united-kingdom-uk-import-value-pharmaceutical-products-from-the-european-union/

'THE WILL OF THE PEOPLE'

I am puzzled by the resistance of many Brexit supporters to a second referendum. Surely this seems sensible. Let's look at the facts.

In 2016 voters in Britain decided (narrowly) to support the withdrawal of the UK from the EU. The Prime Minister and his government were charged with obtaining the best deal possible for the UK and the deal would then be put before Parliament for ratification. In principle, it all sounds straightforward. But as we now know, the truth is very different.

Firstly the Prime Minister who instigated the referendum fell on his sword on declaration of the result. Being committed to remaining in Europe, he felt it was inappropriate for him to lead our exit. So a new Prime Minister, Theresa May, was appointed – one who, although a remainer, was less squeamish about driving through legislation in which she did not believe. Some 20 months later, she has negotiated a deal from the European Union and is ready to put it before Parliament. Parliament is a little less than enthusiastic about the deal on offer and is disinclined to ratify it.

Astonishingly, nobody has bargained for this. The process, as outlined in 2016, doesn't account for the possibility that it might not get through Parliament. The vote to leave the EU seems to have been predicated on some rose tinted view of Britain seated somewhere in the 1950s where we could delude ourselves that we still had an empire and were still a proud island race.

We seem to have been caught completely unawares by the EU's attitude to our leaving. We seem surprised that the EU should wish to make the process unpleasant. I think many who voted to leave in 2016 thought it would be plain sailing and that we would dictate our terms to the EU. After all we were the ones leaving.

From the EU's position it is very different. Obviously they wish to impose as punitive an agreement as they can, if only to discourage other states entertaining such frivolous notions. It is hardly in their interest to be accommodating to our needs. After all, on purely economic grounds alone, the EU needs trade with us much less than we do with them. So we are faced with a "deal" from the EU which is clearly unpalatable to the UK Parliament. The EU has declined to negotiate further and so we are at an impasse, waiting to see who will blink first.

How do we break this deadlock?

Firstly it's obvious that Mrs May threatening the EU with a "no deal" Brexit will not work. That's truly a case of cutting off our nose to spite our face. The EU face can survive the loss of a British nose easily. The UK cannot. A budgetary blip in the EU amounts to an economic apocalypse in the UK. So we need creative thinking.

We need to find out what people think. And what better way to do that than through a referendum.

I hear you groan. And I hear the leave camp dragging out that tired old argument about the people's will. Sure, the people's will in 2016 is well known. But this is 2018, shortly to be 2019. People change their minds. Many who voted to leave in 2016 were naive about what departure from the EU would mean in practical terms. They are rather like a jury who, having reached a decision, will not allow a retrial in the face of new evidence.

In 2016 people voted on the basis of misinformation, misplaced nationalism and a desire to bloody the government. It became a by-election on the government's record rather than a rational considered response to the future of this country.

Two years later many say they would have voted differently. So let's have no more about this being "the will of the people". It may have been then but it probably isn't now. And if we believe the will of the people to be sacrosanct, we should test it again now that we know exactly what leaving the EU will mean.

If a second referendum also comes out in favour of leaving, then this will be a validation of their views. If on the other hand, the 2016 results are overturned, we should be glad that we tested the will of the people once more.

This is much too weighty a matter to leave in the hands of politicians.

2019

ROCK BOTTOM

I retired from the Cure Parkinson's Trust at the end of April 2017. It was time to take a rest and continue my fight against Parkinson's on a more local platform. I honestly felt I had done my bit. Don't get me wrong, there are many who fight all the way but, for me, it was the right decision at the right time. Like an ageing sportsman, I wanted to leave whilst I still had something left in the tank. I wanted people asking why I had perhaps retired early rather than wondering why I was still dragging on.

I'm not sure whether I actually achieved that but people made kind remarks and for the most part it seemed like a good idea. More time to spend with the kids, more time to do all those things I wanted to return to – glass art, photography and so on. Who knows, maybe even a little bit of cricket. Time to enjoy the fruits of retirement while I still could. An opportunity to steal time back from the Parkinson's and to put that time to alternative uses.

It all looked brilliant when I drew up the plan in my head. And for the first few months, it looked pretty brilliant in practice. I would wake up in the morning and ask myself what I wanted to do that day. There were no rules. If the sun was shining, I might take my camera out and do some photography. If the weather was inclement, I could enjoy doing some stained-glass. Or writing. I had always promised myself that I had a novel in me. Maybe now was the time to write it.

But gradually and rather insidiously, opportunity and enthusiasm were somehow transmogrified into indecision and apathy. You can have too much choice. As Bruce Springsteen said of American television it was a case of "57 channels and nothing on". It felt the same to me. Instead of springing out of bed ready for the day's excitement, I was dragging myself up. Household chores usually swiftly dismissed to allow me the whole day of interest soon expanded to fill the time. And if I didn't get dressed at all some days, it was no big deal. And where's the harm in a bit of daytime television.

On the whole I disguised it well. Not even my best friends were aware. Often they were going through problems themselves. And you can only shoulder just so many burdens.

Although I didn't at the time, I recognise it now for what it was and can call it by its name. Depression. For months I rationalised my inertia, mental fogginess, bad temper and anhedonia as a simple product of ageing. A stage I went through. Nothing more. It would all right itself given time I imagined. I didn't look for solutions because mostly I didn't believe there was a problem. And, as any psychiatrist will tell you, getting patients to recognise that something is wrong is the first step on the road to improvement. Sometimes you have to reach rock bottom before the only way truly is up.

Rock bottom for me was one afternoon when I found myself, still in my pyjamas, watching the Jeremy Kyle show. But then of course that would be rock bottom for pretty much anyone. Jeremy Kyle I mean, not the pyjamas.

It was like a slap in the face. I had to do something. Something decisive before I descended into the abyss of daytime television and self-loathing. I started by switching off Jeremy. Silence. No screaming, no punches thrown, no sanctimonious claptrap from the host. Just peace.

At least it was a start.

SPEECH TO THE WORSHIPFUL COMPANY OF BUTCHERS

I received a clear brief for this speech. The brief was 'be brief'. Well being brief is not really what people with

Parkinson's do well. We tend to be rather slow so bear with me.

By any standard, James Parkinson was a remarkable man. We tend to think of him as a medical man but his early interests were elsewhere. He was particularly interested in palaeontology and geology. He was even elected as a fellow of the geological Society and published several books on dinosaur fossils.

Another major interest was politics. He campaigned strongly for universal suffrage for instance, years to decades before it became accepted. And there is some evidence that he was involved in the 1794 attempt to assassinate King George III, which became known as the popgun plot, an attempt to fire a poison dart from a popgun. It's the kind of thing that sounds more like a schoolboy prank than an attempt to bring down the monarchy. Either way it probably cost him his knighthood. Fortunately for all parties he turned away from regicide and back into medicine.

Over the course of his career he published on many aspects of medicine particularly gout and appendicitis but also, perhaps most intriguingly, the correct way of wearing a surgical truss. Mercifully out of print. He even wrote a children's book about dangerous sports.

But by far his most well-known work was "An Essay on the Shaking Palsy" published in 1817.

This was the first work ever to describe what later became known as Parkinson's disease, with its tremor, muscle rigidity, immobility and balance problems. Normally my tremor is not this bad. But then normally, I'm not speaking to 150 livery men. As with most illnesses at that time, treatment consisted of a mixture of toxic potions and leeches. "Take two leeches and call me in the morning".

Fast forward 200 years and the Parkinson's disease we see today is very similar to what James Parkinson himself described. Parkinson himself would have easily recognised the condition although he probably would have wondered where all the leeches had gone. He would probably also wonder where his house had gone – it's now a wine bar. And yes, that probably is the sound of him turning in his grave

So, if I can be serious for a moment, what does it feel like to be diagnosed with Parkinson's?

Nowadays, we have CT scans, MRI images and more blood tests than even the leeches could manage in order to help make as accurate a diagnosis as possible. But in the end it comes down to the same thing. Being diagnosed with Parkinson's is a life changing experience.

Not least because you learn two new words. You learn that it is neurodegenerative which means it will get worse. And you learn that it is incurable. By any standards that is a lot to take in. And for many people, that is as far as it goes. You accept your lot. That is the hand that life has dealt you.

But what if you don't accept your lot? What if you decide to do something about it? What if you question why it is incurable? Many illnesses start out incurable. But medicine somehow finds a way to change that. Polio, smallpox, and nowadays some forms of cancer. Why not Parkinson's disease?

The simple truth was that nobody had asked the question. Everyone had accepted that it was incurable. Until 10 years ago when the Cure Parkinson's Trust was founded.

And it was not founded by scientists. It was not founded by physicians. It was founded by patients. By patients who did not accept their lot in life. Patients who knew that life is not about the hand you're dealt but how you play that hand.

And over the last decade CPT has grown from two people and a typewriter to what it is now – an international charity funding some of the most exciting, innovative and ambitious research into a cure for Parkinson's. It's a charity utterly focused on one objective - to put itself out of business. Above all, it's a charity that never allows people to give up hope.

I have had Parkinson's for twelve years and I've learnt one or two things about how to live with the condition. So I would like to end with some pieces of advice in case you should ever get Parkinson's.

Rule number one – Be careful what you eat.

If you have a tremor, and I think you can see I have, you have to be careful what you eat. Not because of any critical diet requirements, just the mechanics of transferring food from your plate to your mouth. And not over your left shoulder. Or into your right ear.

When we have dinner parties at home, I always tell guests to wear old clothes. Especially they will be seated next to me. Spaghetti is the worst. I pick up the Parmesan cheese and thirty seconds later the table looks like a winter nativity scene. That's the thing with Parkinson's – you run out of clean clothes faster than you run out of hope.

Rule number two – don't go to auction houses.

his is an environment where raising an eyebrow is enough to make a bid. Involuntary movements can get you into big trouble. There is nothing more disconcerting than hearing the hammer go down and realising you have just outbid the National Gallery for a Michelangelo sculpture. So my advice is either don't go at all or get a friend to duck tape you to the chair.

Rule number three – my final piece of advice. Don't make assumptions.

When travelling anywhere these days my walking stick helps get a seat on crowded trains. Most people will offer me their seat. Sometimes I accept but at other times, if I'm feeling well, I will decline. I was on the underground the other day waiting at Southwark station. A train pulled in, the doors opened and, as I stepped aboard, a young man started to get to his feet and gestured to his seat. I felt okay and shook my head in polite refusal. Again he gestured to the seat. It was very kind of him but I made my point by gently pushing him back down and thanking him. I vaguely heard him say something as the train doors closed.

"I'm sorry" I said as the train moved off "I didn't catch that".

"That was my stop" he said.

I would like to say one last thing.

Because I am a neuroscientist as well as a person Parkinson's, I'm often asked whether I really believe in a cure. And my answer is the same today as it was back in 2006. The answer is yes, we will cure it.

There has never been a better time to have Parkinson's.

THE ONLY WAY IS UP

Switching off Jeremy Kyle was probably as decisive a statement as anything I had managed thus far. More to the point, it was a tacit recognition that his sanctimonious custodianship of this toxic cockpit of raw emotion was not where I wanted to be.

But I'm thankful – truly I am – for Mr Kyle's show in one respect. It defined one end of the scale. Rock bottom. Absolute zero. Nobody will ever make a television programme worse than this. But how do you leave rock bottom? How do you climb back into the light? When apathy has, for months, been in the ascendant, how do you break its malign grip?

Lewis Wolpert, emeritus professor of biology at University College London, wrote one of the best tracts on depression I know. Malignant sadness: the anatomy of depression, published in 1999, talks in depth about the complex emotional landscape of depression. Perhaps the defining leitmotif of his book is the sense of shame, the sense that somehow one is weak, lacking in some vital component of one's psyche. And he lays bare the shame when he talks of his grief at his wife's death going on to say that depression was worse.

Let me say immediately that my depression was nowhere near that of his malignant sadness. I call it depression because, on self-diagnosis, it met the clinical criteria with a MADRS of nineteen and a Ham-D score of fourteen. Depression is a range of depth, from mild apathy and anhedonia through to an all-consuming sense of worthlessness. I was very much in the shallows of depression rather than its black depths, where the wild things are. But it gave me a glimpse of what kind of place that was, and where that whirlpool of my darkest imaginings could take me.

There are of course different ways of looking at things. And I won't insult your intelligence with the trite glass-half-full-half-empty metaphor. But the way in which we recognise depression has significant bearing on how we choose to address it. If we ascribe greater intensity to our experience than is the case, we medicalise the experience immediately – it calls for drugs. If on the other hand we play down what our bodies are telling us, we may underestimate its consequences.

My background is in neuropharmacology, the action of drugs on the brain. For many years I headed a team looking at neuropsychopharmacological aspects of brain function. I can even remember some of it!

I should nail my colours to the mast right here. There are plenty of people who see antidepressants as part of a grand conspiracy theory by the drug companies to sell us medicines at whatever cost. I do not see it that way. My view is much more simplistic. I see depression as a neurochemical imbalance, in the same way that diabetes is a neurohormonal imbalance. It is only our stigmatisation of the one but not the other that allows people to draw such fanciful conclusions. If we stigmatise depression and further stigmatise antidepressants, we do a great disservice to those who experience depression and struggle with the choices they make.

Don't get me wrong. I am as critical of the pharmaceutical industry and its motivations as anyone. And that is probably a subject for another day.

So my first decision – well, second after switching off Jeremy Kyle – was to decide whether I needed medication. And you may be surprised to hear, especially after the above, that I chose not to seek medication. And that decision was taken not because of any inherent reticence over drugs but simply a desire to see what could be achieved without.

So, no drugs. Where do we go from here?

MIND GAMES

OK, look this one up.

Roger Koenig-Robert et al. Decoding the contents and strength of imagery before volitional engagement, Scientific Reports (2019). DOI: 10.1038/s41598-019-39813-y

I'll get straight to the point. This is an important paper, albeit one hiding behind the most reticent and least illuminating title. Let me translate it for you. In simple terms the scientists asked people to imagine one of two images. The choice was theirs. When they had chosen, they immediately pressed a button. Meanwhile the scientists looked at their brain activity. To cut a long story short, the patterns of brain activity predicted what the choices would then be. And these patterns occurred before the subject made their conscious decision. In some cases, up to eleven seconds before they decided. In other words, the scientists knew what decision the subjects were going to take before they took that decision.

So what does that mean? Firstly it means that it is possible, without being sensationalist, to read the thoughts of the subject, albeit within a limited range, say the choice of red or green objects. But I'm sure it won't be long before this limited capability becomes a much more extensive facility. CCTV already knows where we are. FMRI can now say what we are thinking. More to the point it can tell what we're thinking before we can.

Secondly, it suggests that conscious decision-making is an illusion. The brain makes the decision and conveys this to the mind. The brain allows the mind to continue to believe that it, the mind, makes the decisions. The opposite is true. In other words, consciousness is a narrative rather than an executive function. It reports, in the first person, what it has received in the third person from the brain. The mind is simply the way the brain presents itself the outside world.

Most of our legal system is of course based on personal responsibility. In other words, we are responsible for our decisions. That's all well and good if we (and by "we" I mean our minds) are making decisions. But if we are simply the product of an internal decision-making system, where does that leave us? How can one be culpable if one is not responsible? If the brain is taking the decisions based on its own agenda, how can we blame the mind, acting as a go-between.

Of course it's not just legal issues that are relevant. If you take away the notion that the mind decides, you instantly take away free will. Most religion is based on the notion of good and evil and a succession of choices between those two poles. If we have no free will, we are innocent of our choices and their consequences. In catholic terms, we are free of original sin.

Sounds good? Well, not really. Take away free will and accountability and you lose track of the notion of karma. No free will, no karma. In other words the baddies go unpunished. I could go on. Of course, those of a religious persuasion can always argue that, whatever the data, the

Almighty would certainly have the wherewithal to cover his tracks. It's not my place (or intention) to offer an opinion here.

The authors of the paper couch their findings in terms of post-traumatic stress disorder and the ways in which images filter up from the subconscious to the conscious. They brush aside the notion of volition in their concluding remarks – not as a dismissal but more as the recognition that they have opened a pretty big can of worms.

Food for thought? You'd have to ask the brain.

GDNF – WHERE NOW?

It has been an interesting few weeks. On 26th and 27th February, the results of the double-blind, placebo-controlled trial of GDNF in Parkinson's and of the open-label extension study were published in Brain and Journal of Parkinson's disease respectively. The following day the first part of a double bill documentary on the trial was broadcast in the UK on BBC2. The following week the concluding part was also broadcast.

Between the papers and the documentary, the subject has been thoroughly examined and dissected. Discussion groups have been alive with interest, comment and interpretation. Many comments have been positive and congratulatory to the study participants, designers, scientists, documentary makers and the charities that funded the trial. Some comments have been negative, often expressing frustrations driven by the dichotomy between documentary title and final conclusions. Some thought that words like "miracle cure", even when followed by a question mark, misled people over the content. If I'm honest, I would agree.

None of that should take away from the imagination of the scientists who devised this trial and from the naked courage of those who participated in the study. Let's be under no illusions here – these men and women put their lives on the line by being part of the study. Alan Whone, the principal investigator, said as much at the beginning of the documentary. This was experimental surgery, an experimental method, and an experimental drug. The possibility of catastrophe was very real. And that's not hyperbole. It's fact.

Of course the documentary made much of the bravery of the participants (and rightly so, in my view), putting a human face on the raw numbers of science. Emotions ran high and low. A rollercoaster, if you must use that tired tabloid simile. And nowhere were the emotions more raw than in the sequence describing Tom's death. Tragic – almost comically tragic – that the person so central to the instigation of the trial should be the only one unable to benefit. You could almost hear Tom laughing at the injustice of it.

So what are we left with? What is Tom's legacy? If we strip away, for a moment, the human dimension and look simply at the numbers, what does the data show? Did GDNF work or did it not?

To answer that, you have to move away from the documentary and go to the papers themselves. The only piece of data presented as such in the documentary was the moment when Alan Whone revealed, in what was uncomfortably close to game show host mode, that the drug had failed to reach its primary endpoint.

But what does that mean?

To answer that question, you need to know a little bit about clinical trial design. Bear with me – I'll try to keep this as short as possible.

Basically all clinical trials have what's known as a primary endpoint. This is essentially a clinical measure chosen before the trial begins to determine efficacy over placebo. Depending on the nature of the trial, the primary endpoint can be one of many things. In some studies for instance, quality of life may be the main interest and a quality-of-life measure would therefore be chosen as the primary endpoint. In a cancer trial, the main interest might be duration of survival and an appropriate endpoint would be chosen to reflect that. But for trials of Parkinson's drugs, the norm has been a measure called UPDRS (Universal Parkinson's Disease Rating Scale) or at least the motor component of the scale.

The study investigators set themselves the target of a 20% improvement in "off" state motor UPDRS. GDNF failed to reach that target. Nor did it show any difference on secondary points. End of story. End of GDNF.

Hang on a moment. Not so fast. When you look at the post hoc analyses (the ones that weren't included at the beginning), a rather different picture emerges. 43% of patients receiving GDNF showed clinically important motor improvement. None receiving placebo did. PET scans of patients receiving GDNF showed between 25% and hundred percent increased fluorodopa uptake in the putamen. Not one placebo patient showed a change.

These results are a little more difficult to reconcile with the failure to demonstrate effectiveness at the primary endpoint. But the rules of statistics are such that one needs to define one's endpoints in advance of the trial not at the end. If the post hoc analyses had been included at the beginning, we might be talking about the data slightly differently. Imagine for instance if fluorodopa uptake had been chosen as a primary endpoint. The study would have been a triumph. They would be singing and dancing.

But that's the cruel world of statistics. A beautiful hypothesis apparently slain by ugly fact. And of course you are welcome to believe that should you so wish. But I think you would be selling the study short if you did.

It is the easiest thing in the world to utter the standard scientific mantra that "further studies are required" before conclusions can be drawn. The general public, where science is concerned, is not fond of cliff-hangers. I'm a patient and a scientist. As a patient, I want answers. As a scientist, I want to know that the answer is the correct one.

How do you marry up those two aspects? How do we blend the patient's sense of urgency with the scientists need for accuracy? The longer you look at it the more obvious is the answer. You have to involve patients. Study designs have to reflect patient experience. And, more than that, that experience needs to be central not peripheral.

"GDNF works, I know it works" said Steven Gill "it's just a matter of getting enough of it in the right place". We've learned a lot from this trial. And we owe a lot to the forty one souls who put their lives on the line in this trial.

To paraphrase Churchill "Never in the field of human neurosurgery has so much been owed by so many to so few".

A LAUGHINGSTOCK

You have to laugh really. Since the Brexit referendum in June 2016, we have had nearly three years in which to negotiate a rational, fair and acceptable deal that would enable us to leave the European Union with dignity and above all with maintained strong trading links. Somehow, whether through wilful misunderstanding or plain and simple incompetence we have managed to antagonise more or less the entire European Union, most of our own electorate and even the various political factions. We have somehow managed to turn a clear (if fatally flawed) ideology into a national humiliation.

Britain, even at its most eccentric, has always consoled itself with the knowledge that, no matter how bad things got in Britain, we would at least never be more embarrassing than the United States under the present administration.

We had nothing to touch Donald Trump for hubris and moral vacuum. Somehow that was strangely consoling.

Now we have nothing. Our country, the once proud imperialist nation, is now an international laughingstock. Whereas once being British was winning first place in the lottery of life, we now hold the wooden spoon. We are a nation ill at ease, split so many ways and floundering. We are, to quote the famous lines "not waving but drowning".

Our politicians on both sides are playing a terrifying game of brinkmanship. Mrs May (at time of writing still the Prime Minister) clearly plans to take the decision on her deal down to the wire. To the point where no alternative is available but to crash out of Europe in a blitzkrieg of acrimony and litigation. This is not how politics used to be run. This is not the sensible politics of consensus. This is bullying.

Perhaps the saddest part is that the hard line Brexit supporters actually support a no deal Brexit. In some distorted notion of patriotism, they feel that this is the best way to send a message to Europe. They are certainly correct – it sends the message that we are prepared to sacrifice our own economy to make some fatuous Little England statement. No wonder they are laughing at us.

But this is no laughing matter. To see the Brexit crowd giving audience to Tommy Robinson, odious leader of the neofascist English Defence League was to witness rabble-rousing of the worst kind. It was hard to believe this was London in 2019. It felt like Nuremberg in 1934. Finally we were seeing Brexit for what it is, stripped of its pleases and thank yous. An ugly plebiscite.

I'm tired of hearing about "the will of the people ". Firstly, the original vote was, at best, a marginal victory for the leave campaign. Anyone who argues that 52% versus 48% is "a mandate" is living in cloud cuckoo land. It is, to all intents and purposes, a coin toss. And it's funny also that those who squeal loudest about the perversion of democracy are those who would deny the electorate the opportunity to have the final word in a second referendum. Apparently the electorate is not allowed to change its mind. I wonder what they believe general elections are about.

But if we believe in true democracy and the upholding of the will of the people, the case for a second referendum is clear. The first referendum set out the chosen course of action. The government has attempted to implement that. The second referendum would be an opportunity to endorse the government's work. Or not as the case may be. It has, after all, been the best part of three years. Plenty has changed. Lies and misrepresentations have been exposed.

To deny the population of the UK that right on the grounds of democracy is comical. For the Brexit supporters, democracy apparently only applies once. Make no mistake - those who would deny us a second referendum, those angry, tattooed, shaven headed bigots in Parliament Square on Friday, are the true enemies of democracy.

And in the midst of this chaos and anarchy, our political leaders are immobile, fixed in the headlights. The Prime Minister is clearly fatally wounded, damaged by her own hubris and further undermined by her own colleagues. This is understandable. The Tories have always been savages beneath the skin. And they can't help themselves when there is blood in the water.

Instead of uniting, our politicians seem hell bent on furthering their own personal agendas whilst still muttering about the public's loss of faith in democracy. I don't think the public has ever lost its faith in true democracy. Where credibility has been stretched beyond breaking point is our faith in our elected representatives as bastions of said democracy. Instead they are vultures picking over the carcass of accountability.

It's hard to know where we go from here. How do we step back from the brink? I truly believe we are teetering on the edge of Civil War.

NO CAUSE FOR CELEBRATION

Let me give it to you straight. I don't celebrate World Parkinson's Day. I don't celebrate World Parkinson's Month. I don't participate in discussions on whether it's Parkinson's or Parkinson. I will lose no sleep over the decision to drop the word "disease" from its name. I don't care whether we call ourselves Parkies, people with Parkinson's (PWP) or people living with Parkinson's (PLWP). And I certainly don't make a point of marking my Parkiversary, the anniversary of my formal diagnosis, and surely the most ridiculous of causes for any kind of celebration.

I don't think there is very much about this condition worth celebrating. World Parkinson's Day on 11 April, is the anniversary of James Parkinson's birth. And every year we stand with our wan smiles for the camera, vowing to be a warrior, a soldier, or whatever. We tell each other that "I may have Parkinson's, but Parkinson's does not have me".

Oh but it does. It so does.

Each year, World Parkinson's Day reminds us that it still exists. It reminds us that billions of dollars have been spent on research. It reminds us that billions more dollars are still needed. It reminds us that, since World Parkinson's Day last year, thousands have been told that they have Parkinson's. And tens of thousands no longer have Parkinson's. Because they're dead.

Last year, Parkinson's was an incurable disease. This year, Parkinson's is an incurable disease. And let's recognise the fact that until that day when we win, we lose. Every day, we lose a little bit more.

And every year, on 11 April, we remind the world that we haven't gone away. Which is of course a nonsense. Because many of us have. The condition itself certainly hasn't gone away. Each year, on this one day, we draw the public's attention to this condition and the need for more treatments, better treatments and even perhaps that final treatment we all dream about.

That's the difference between people with Parkinson's and the rest of the world. You are only reminded once a year. We are reminded – and boy are we reminded – for the whole year. So, no I will not be celebrating World Parkinson's Day. I would just like one day every year – is that too much to ask – when I don't have to think about this condition.

One day we will beat this thing. In the face of everything, it is hard to keep saying that. But it's true and I believe it now as much as I ever have. And when that day comes, when we are finally delivered from this hateful condition, I will dance on its grave.

Then, and only then, will I celebrate World Parkinson's Day.

PS: If you want to bring that day closer, please feel free to donate to The Cure Parkinson's Trust

CONFERENCE BAGS

I've been going to scientific meetings, congresses and conferences, national and international, since 1982. That's 37 years, all told. And in that time I've probably been a participant at every kind of scientific meeting you can imagine. I've attended (and delivered for that matter) plenary sessions, workshops, think tanks, poster sessions, breakout groups, brainstorms, advisory panels, executive committees, you name it. I've heard presentations as short as five minutes or as long as two hours (both scheduled for an hour, the former reducing the session chair to 55 minutes of embarrassed coughs while scanning the audience for any movements that might portend a question.

In the latter case, the speaker delivered a sort of free-form jazz improvisation kind of lecture, laid-back and self-indulgent when it should have been incisive and punchy. When this eventually ground to a halt, the session chair intervened briskly to douse any remaining embers that might be coaxed into life. Sad to say but the speaker was a Nobel laureate and should have known better. And no, I'm not going to name him. Or her.

Some meetings have been small enough to fit in my living room and, before you comment, I do not live in a Saxon castle or Tudor baronial manor. My living room is little more than a couple of sofas and the dog occupies one of those, growling if people try to dislodge him. Hardly a major conference venue. I can't even promise any catering. At a pinch, I can probably run to some chocolate hobnobs. And I'm pretty sure I saw some fig rolls at the back of the cupboard the other day.

Other meetings have been huge, filling venues the size of football stadia. Each year the Society for Neuroscience draws some 35,000 neuroscientists to one or other American city. In truth, only a tiny handful of cities, even in America, can accommodate what amounts to an entire army. Robert E Lee's Army of Northern Virginia was smaller when he crossed the Potomac and headed to Washington. The logistics of conferences on this scale make the Normandy landings look like a fraternity beach party.

I've come to realise that one thing is common to all conferences, big or small. I'm talking, of course, about the conference bag. You know, that absurd sack emblazoned with an embarrassingly colourful logo that, no matter how hard you try, you cannot seem to hide. And bear in mind that there may be tens of thousands of these, all identical, within the area of the conference centre. It's only a matter of time before you accidentally pick up somebody else's.

I'll let you into my dark secret. I used to collect conference bags. There, I said it. I've been to over a hundred conferences over the years and have brought home conference bags from most of those.

Why, I hear you all ask. Well, those of you that aren't already scrolling down. I think it's because they are so varied. In size, shape, number of straps, size and number of compartments, organisation, colour, materials and so on. There seems to be no unified or agreed standard on what makes the definitive conference bag. Some are almost like carrier bags. Others are highly compartmentalised, like Oriental medicine cabinets. But, like fingerprints, no two conference bags are the same.

Back in the 80s, the conference bag had very simple criteria – it had to be large enough to accommodate the program and abstract booklet for the meeting, along with the usual handful of flyers for scientific instruments, more conferences in the Far East, a biro that never worked and a notepad. Usually your conference name badge was in there as well, often with the safety pin open, ready to draw blood. And that was it. The flyers and other promotional material went straight into the first bin you found on your stroll round the exhibition hall looking for freebies.

Over the years the conference bag evolved. By the mid-90s there were little pouches for your mobile phone, a larger zipped section for your laptop and the usual loops to hold biros, still unreliable. Gone were the notepad (who wrote notes on paper anymore when you had laptops) and the obligatory CD-ROM of the proceedings and abstracts.

By the time we were comfortably into the third millennium, there was no need for a mobile phone pouch. The cell phone was the size of a crispbread and fitted comfortably in your trouser pocket. You didn't even need to wear the special trousers any more. The large laptop section was replaced by a tiny pochette for your iPad and nobody used Biros any more. As for the CD, it was now in some elephant graveyard of redundant apparatus, replaced by a web link. The conference bag was smaller and smarter.

Even the materials have changed. Gone are the leatherette briefcases that smelt like the interior of the

1960s Cortina. Out too are the girly over-the-shoulder tote bags (really, who thought that was a good idea?). Minimalist seems to be the idea now. Small and minimalist. And made of biodegradable materials. The WPC bag for 2016 was essentially a small hessian handbag. A neat design and popular with all. Perhaps it's only design flaw was the incredibly fiddly and unhelpful clasp. Bear in mind that more than a thousand of the conference attendees had Parkinson's, with significantly compromised dexterity. Still it kept the fingers working. Often for hours.

I have thrown out nearly all of my conference bag archive. The Science Museum have stopped answering my phone calls. So the whole lot went in the bin and will doubtless, in centuries to come, be rediscovered in some piece of landfill and puzzle archaeologists of the time.

I have kept one. Only one. But it's a special one. From the 2007 ECNP (European College of Neuro Psychopharmacology) meeting in Vienna. A thing of beauty and design brilliance, it is the Swiss Army knife of conference bags. It has pouches for everything from (working) Biros through to changes of underwear. Space for iPads, kneepads and keypads. And the whole thing falls into a neat backpack. I use it as a mobile office – with pens, paper, staplers, hole punches, a paper shredder and a photocopier.

Okay, I lied about the last two.

My bad.

KYOTO DIARY 1: PREPARATION IS EVERYTHING.

I like to delude myself that I am organised, that I approach every trip abroad with calm and poise. In my mind I am a Zen master. The very epitome of organisation. In the zone.

In fact nothing could be further from the truth. It has taken until today, only three days before the first salvoes of WPC 2019 are fired, for me to realise that entering Japan is not the same as nipping over the channel to pick up a few bottles of cheap burgundy. There are preparations to be made. Complex preparations. And ones that should really not have been left until day K-3.

This realisation has transformed me from a Zen master to a gibbering, disorganised halfwit in a matter of an hour or so. It is one thing finding that you don't have enough shirts washed. It is quite another to realise that you have insufficient medication for the trip. And Japan is not the kind of place where you just turn up at the pharmacy and say "Twenty tablets of your finest levodopa if you please, squire. Quick as you can". Firstly, they will not understand you. And secondly, they will almost certainly ignore you.

The voice in my head is telling me, in increasingly strident tone, "You should have thought of this earlier, dimwit. And by the way, you don't have enough shirts either". So this morning has been spent pleading with my GP to write a prescription on the spot (and if you really want to antagonise a GP, this is a brilliant way of doing it). Then a swift dash (figurative not real - obviously) to the pharmacy, all the while praying that they will have everything in stock. As it happens, they do. Today the gods of pharmacy smile and they even make up the prescription in real-time before they close for lunch after my pleading (and if you really want to antagonise a pharmacist, this is a brilliant way of doing it).

So lunchtime arrives and my stocks of medication are now replenished. Which is more than can be said for my stocks of goodwill at the surgery and pharmacy. Normally the polite thank you letter does the trick, reminding them that they are appreciated. Which they are. But this has happened too often for them to be so easily placated. I need to up my game.

So the first crisis is averted. What next? Oh yes, I remember – shirts.

For some reason, my shirts are not where they should be and, only after an hour of brow furrowing do I remember their location. In a suitcase in the loft. Last September, as summer ended, I decided to create more space in my wardrobe by putting all my short-sleeved shirts and T-shirts into suitcases in the loft, safe and secure until next spring. Logical, sensible and space-saving. I remember congratulating myself on my brilliance.

However...

While I remembered the day in September, I had forgotten the day last November when we, as a family, decided to throw out all the old suitcases we no longer needed. It's my own fault really. Before I remembered their contents, they were already in the hands of the charity shops and presumably winging their way to places of urgent need. Pretty much my entire summer wardrobe. It's strange to think that somewhere in Ethiopia or Eritrea there are now people dressed exactly like me. They probably think the same.

Strangely charming though this notion is, it doesn't solve my immediate problem – it will be 30°C in Kyoto and my stock of winter woollens are hardly going to cut the mustard. Summer shirts are needed, and urgently. This is a job for Amazon Prime. Within 20 minutes I have ordered an entirely new wardrobe. Five shirts of linen and cotton, made in Nepal and Bangladesh. Some day they will doubtless find their way to some war-torn part of Africa but, for the moment at least, they are in my suitcase. Part of my suitcase will always be Kathmandu.

Medications for the trip arranged, located and purchased – TICK.

All summer clothes despatched unknowingly to Eritrea - TICK

New shirts purchased from Kathmandu and Dhaka - TICK

I am once again the Zen master. All is calm. I breathe in ... I breathe out .... And relax.

Suddenly I am woken from my reverie, my body cold and sweaty – where did I leave the passport?

No, seriously, where did I leave it?

KYOTO DIARY 2: ONE MORE SLEEP

Really feels like that. Suddenly I'm a 10-year-old boy, waiting for Christmas to arrive – not quite sure whether Santa Claus exists or not but just happy to enjoy the moment. And it's all come round much quicker than I expected. One minute WPC is an abstract concept, somewhere over the horizon. Then suddenly it's on our doorsteps, the fruition of years of planning by a whole bundle of people.

Of course when I say "whole bundle of people", that's true but every part of that bundle would acknowledge the driving force behind WPC. Formally, she is Elizabeth Pollard, Executive Director of the World Parkinson Coalition. But to those who know her (and to know her is to love her) she is simply Eli. She is the person who makes WPC happen. She is a cog in every clock, the hub in every wheel. Eli knows everybody. Not just everybody who is anybody but everybody, full stop. Or everybody, period, as you persist in saying in North America! When you see her next week, take a moment to say thank you.

Of course it's not really one more sleep until WPC. It's one more sleep before I get on an aeroplane for that 12 hour trip to Osaka. Then by land to Kyoto. I'm about as prepared as I am ever going to be. And yes, thank you for asking, I did find my passport. Inevitably it was in my "safe place". This of course questions the wisdom of having a safe place if you can't remember where it is or even, as in my case, the fact this one even had a safe place. How many times have you tied a knot in a handkerchief to remind you of something, only to be bewildered later to find this knotted handkerchief. I often think of my declining memory as 1000 tiny rodents gnawing away at the wires of my cognition.

I'm not a great flyer. My Parkinson's has reached the stage where restless legs plague me most of the time and especially so when constrained to sit still. Such as in a concert or on an aeroplane. So 12 hours of sitting on my hands or writhing in my seat is a pretty depressing prospect. That said, it probably won't be a million laughs for the poor soul in the seat next to me. I shall do my usual thing of apologising, explaining that I have Parkinson's telling him or her a little bit about it. And they will do their usual thing of being sympathetic, polite and kind. But it still won't assuage their anxiety as the involuntary jerks narrowly miss knocking over that glass of red wine. I don't yet know who you are and nor do you. But let me apologise in advance for the most miserable flight you are likely to experience in your life. Just send the dry-cleaning bills to me.

Of course, many of the more sensible advocates are already in Kyoto and will be well acclimatised. Already there have been posts from most of Kyoto's attractions (and there are many). "Wish you were here" they say. "Me too" I think. There is a palpable and rising sense of excitement.

But there is also a sense of duty. Not everyone who wanted to go to Kyoto could. It's a long way away and a lot of money. So each of us who can attend needs to make the best of it, not just for ourselves but for those who can't attend. We should write stuff down, take notes, make observations and do our best to make sure that we can pass our learning on. So yes, I am excited about going to Japan. Yes, I hope to learn a lot. And most of all, yes, I will try to pass it on.

It's nearly 11 PM and I have everything collected together on the dining room table ready for tomorrow's departure. I just need to pack the suitcase now. Tomorrow will be a great leap into the unknown. But today at least is defined by certainty – the certainty that I will have forgotten my toothbrush. There are always some things you can rely on.

KYOTO DIARY 3: ON OUR WAY

5am British Summer Time (BST). Predictably I woke around 5am, as usual and in pain. Not my choice you understand, just a reflection of my general insomnia and one of my increasingly frequent nocturnal brawls - surely among the more tiresome manifestations of RBD (Rem sleep Behavioural Disorder). Still, it gave me an opportunity to download some recent review articles. I am co-chairing a couple of sessions at WPC and wanted to bring myself up to speed on the subject matter. There are few feelings worse than looking clueless in front of several hundred people bristling with questions like a porcupine on crack.

7am BST. Lady G emerges for coffee as I finally wrestle the two halves of my suitcase together. It is like a giant clam and I am a pool of sweat. Lady G asks if she can put some more things in the clam. I greet this enquiry with a look that would curdle milk.

9.17am BST. Our driver arrives and whisks us to Heathrow in double quick time. We are ridiculously early so Lady G fetches espresso while Whatsapp pings away as, one by one, the team gathers.

11am BST. We are all assembled - Lady G, Eros, Tom, Helen, Clare, Vicki, Leah, Rachel, Phil, Clare, David, Jordan, Richard and Kerry - a phalanx of wheelchairs and walking sticks rumbling our way through security, fumbling with our belts and boots, all scanned, swiped and swabbed.

2.30pm BST. After a brief delay, we are finally airborne. Rachel, Clare, Gaynor and Vicki look like the Spice Girls with their matching pink "Parkie Girl" sweatshirts. The Japanese are bewildered as the girls pose for photos. So are we.

3.40pm BST. You've got to laugh - the in-flight entertainment is showing the BBC2 documentary about the CPT/Parkinson's UK trial of GDNF. Vicki, one of the trial subjects and a star of the documentary is sleeping, two seats away. I briefly flirt with the idea of waking her for an autograph.

8.00pm BST, 4.00am Japan time. Over Tomsk in Russia, heading towards Kyzyl in Tuva. Still three thousand miles and more than five hours away from Osaka. Can't sleep. Legs restless. Spasms of dystonia. Medication barely touching it. Really struggling. Then, out of nowhere, Eros appears with a basket of mint Magnums he has creatively relocated from the galley. He is suddenly very popular. He may be less popular when the cabin staff find them missing.

5.51am Japan time. Over Mongolia, heading towards Ulaanbaatar (wow - a place name with five 'a's). Two thousand and fifty nine miles to go. Can't decide whether to take my night-time (UK) meds or start a new day on Japanese time. Still haven't slept. My iPod has kept me going with Joni Mitchell, the Be Good Tanyas, Tom Waits and Van Morrison mostly. Running on empty now (me, that is. Thankfully not the iPod).

8.09 Japan time. Over the Yellow Sea, heading towards Seoul. 80 minutes to landing. Cabin lights on. Breakfast imminent.

9.37 Japan time. Landed at Osaka with an almighty crash as the wheels hit the runway. Bits of decor fell down. Even the stewardesses winced. Welcome to Japan.

KYOTO DIARY 4: ZOMBIE APOCALYPSE

Immigration in Osaka was a weird and dislocating experience. A dozen of us advanced collectively on a determined but hapless immigration official – he insisted on scanning all our baggage while we swept slowly forwards like Hokusai's wave. Eventually he capitulated as we swept forward.

Customs officials don't like sunglasses. Or more specifically, they don't like you wearing sunglasses. It's all about eye contact I believe. I have brought a book of etiquette with me for this trip. The Japanese are intensely polite people. We British like to think of ourselves as polite and civil but, compared to the average Japanese, we are ill mannered ruffians. I think some of the youngsters find this politesse quaint, archaic at best. I disagree. I have taken great pleasure in the exaggerated head dips and bows. Even my own feeble efforts to mirror this politeness are greeted with beaming smiles. I think we should have more of this.

What I definitely need more of is sleep. I did not sleep on the aeroplane. So I went through customs slightly discombobulated. I would certainly sleep following night, right?

Wrong. Much of the night was spent walking the corridors of the hotel. I even posted a Facebook message pleading for cookies or sleeping tablets. Several sympathised but were unable to help. Ben, good friend that he is, offered a fine slab of chocolate in exchange for my wisdom on the coffee maker. I think I got the best of the deal. As I munched my way through the chocolate, with all the refinement of a velociraptor, Ben attempted to get any kind of brown liquid from this spitting Hellcat masquerading as a coffee maker. We talked about skin grafts as it became clear Ben was losing the fight.

So I was slightly discombobulated when we arrived in Japan. By mid morning, I could hardly move. The meds were not working. Jean summoned a wheelchair and had one of the volunteers push me back to the hotel. I hate wheelchairs at the best of times but to be amongst the earliest casualties of Congress seemed particularly embarrassing.

By the time of the congress opening ceremony, I barely knew who I was. I couldn't remember the name of my hotel, or the names of my children for that matter. I could not count forwards or backwards and quivered like a demented blancmange. I would've scored probably 12 on an MMSE. My speech was slurred, my thinking foggy and my strength barely able to keep me standing. I was a zombie. A very well-dressed zombie perhaps but a zombie nonetheless. Kerry, a saint, fetched me food and wine while I attempted to string sentences together in a credible facsimile of conversation. The rest of the CPT team began to discuss evening plans. These already sounded ambitious. By the time Heather arrived, with mutterings of late night karaoke, I waved the white flag and headed to the security of my bed.

And guess what – I slept! And now, like Arnie, I'm back. Normal service is resumed. And just-in-time – I have a poster at lunchtime and I am co-chairing a session in the afternoon. And, you know what – I'm ready.

Bring it on

KYOTO DIARY 5: GETTING DOWN TO BUSINESS

The lights finally came on, so to speak, this morning. Somehow the previous twenty four hours have been conducted entirely on autopilot. I can remember hardly anything. But apparently I moderated a speaker session. And not just that – the session was punctuated by an entirely impromptu five minute dance and exercise session. Seriously. I kid you not. The Japanese clearly have a different way of doing things.

I'm told I also delivered a thirty slide presentation during the day. It even received a rating of four out of five which is impressive bearing in mind I have no recollection of doing the presentation.

Today was different. Today was the first day of the conference proper. Three thousand attendees I'm told. Met plenty of old friends and quite a few new ones. It's always like that on the first day. You're so busy seeing people again that everybody is greeted with "Hi. How are you doing? Got to dash. Catch you later." The irony is that, although it's the first thing you say to each friend, it's often also the last, because you never seem to find that person again over the course of the three days. You are never in exactly the same place at the same time. And in any case, suddenly you can't remember their surname anyway. Blame that third Sapporo last night.

Today I woke up after four hours sleep, ready to rock and roll. My brain, even after this tiny morsel of dopamine was ready to spend it wisely. My poster was presented and the session I chaired attracted around four hundred people. A decent house for a Wednesday. Especially considering the subject matter – "Depression, apathy and anxiety"

I made a brief bid for "Comic Moment of the Day" by falling asleep during parallel session 3 (Sleep Disorders). Normally I get away with this kind of thing unnoticed but apparently this time I had been snoring. In any case, my iPad and iPhone hit the ground at the same time as me. I hastily got to my feet with a sort of yes – I – really – meant – to – do - that sort of air. Still if you're going to do something like that, do it in style.

But the true Top Comic Moment, for sheer surreal madness, goes to the Clinical Research Village. In the middle of a panel discussion about cultural differences in research styles, they were interrupted by a band of Taiko drummers. Well I say interrupted, but that's the point – they were determined to carry on despite the noise. And if you have ever heard Taiko drums, you will know that one is loud, ten practically deafening. Panel session versus drum ensemble... There's only ever going to be one winner.

Like I said, the Japanese have a different way of doing things

KYOTO DIARY 6: A LITTLE NIGHT MUSIC

Let me nail my colours to the mast. I come from a family of music makers, some orchestral musicians, even opera stars. I have listened to countless children's recorder ensembles, heard violins scream for mercy and wince as youngsters try to choke the bagpipes. So I know what good music sounds like. It is the sound of joy, pleasure in making music and pleasure in listening with open ears and mind.

Last night's WPC musical soiree was wonderful. Sure there were teething troubles. Perhaps the microphones weren't loud enough. Maybe the repertoire was unfamiliar or alien. Maybe some preferred to carp about the number of seats. Perhaps all the chilli shrimp had gone. Or they had to drink white instead of red. But at the end of the day, music is about pleasure, not the accuracy of notes on the page. And there were some wonderful moments.

Everyone has their own personal favourites. I loved Eros's splendidly kitsch version of Spandau Ballet's anthem Gold. Tom's sopranos sax was wonderful too in what I think was it an old Herbie Hancock tune. And Emma had them creased with her stand-up routine. And then there was Omotola's gorgeously husky rendition of the cup song. It got the hairs on the back of my neck up.

It was music that made you tingle. And the whole thing was held together by Gaynor and Brian, in matching gold outfits trading banter. Pure fun. Pure joy.

And I even joined those on stage to sing John Lennon's 'Let it be'.

There was a lot of music and a lot of love. The Parkie family is still the house of love. There will always be Statler and Waldorf, waiting in the wings. But then, if you haven't got it, maybe you don't get it.

KYOTO DIARY 7: CODA

It's Saturday morning. Early Saturday morning and we are heading to Osaka on a small bus. WPC 2019 is over. The exhibition hall which, only yesterday, was a hubbub of noise and Japanese drummers, is silent and dark.   The exhibition stands have been torn down, the company reps, cell phones buzzing, are heading to the next conference. The monsoon season has begun and the raindrops fall heavily on the azaleas.

The mood on the bus is reflective. This conference has taken its toll in so many ways. The team is returning bloodied but unbowed.   Battle-weary but battle-hardened. Overtired, overemotional, overdrawn. There has been no shouting or hissy fits. No tantrums, no divas. We're not like that. No, this is just the raw emotion of people who grasp only too clearly the enormity of the task ahead of them and the part they have chosen, or that has been chosen for them, in what we have to do.

It is a noble, even heroic, endeavour – to cure the world of a disease, to wipe it off the planet, to erase its existence. To change the world in such a way that nobody will ever have to suffer from this disease again.

As the bus rattles on to Osaka airport through the morning mist, conversation is sporadic. Most are caught up in their own thoughts. Maybe dreaming of home. It's too early to say where Kyoto fits alongside Washington, Glasgow, Montréal, and Portland. Too early for that kind of perspective. There is plenty of time for that. Certainly we have learnt new things

What was the worst thing about Kyoto 2019? Trolls. Voices of whingeing negativity. We don't have a cure but we are trying. God we're trying. You would think that would be enough. But not for the trolls. They too are trying. Very trying.

Every bold advance brings them out - ugly nihilistic trolls scuttling among the rocks, peddling petulance and pedantry. Left unopposed, they sap the spirit, wither the will. These are the kind of people who would complain about the cooking at the Last Supper. And you know the most amazing thing? Sometimes they are not even people with Parkinson's.

And the best things about Kyoto 2019? The dogged persistence of the battle. We were never part of the It-will-be-over-by-Christmas brigade. We always knew it would be longer. Gone are the glory boys, the headline seekers. Gone are the show ponies. What's left are the fighters. Those who are in it for the cure not for the headlines. These are the real soldiers. These are the men and women I want to stand beside. These are the troops who will fight to the end.

We need to remember that progress is incremental not explosive. The treatments we have today are better than five years ago. And those then better than ten years back. While we lick our wounds, the bus rattles on.

As the rain falls in Kyoto we remind ourselves that, although the job is unfinished, there has never been a better time to have Parkinson's. And it's going to get better.

• Dedicated to the memory of Ryan Tripp, a true soldier and a very good friend.

HOTEL

I've just come back from Japan having spent several exhausting days in Kyoto at the 5th World Parkinson Congress. The conference centre is on the edges of Kyoto, not quite in the middle of nowhere but certainly detached from the main centre of the city. Attached to the conference centre is one of the more unusual hotels one might encounter. A large oval structure on several floors, rather like a football stadium with rooms. I gather it was built only recently. Surprising considering the place had a vaguely Art Deco feel about it. Plush velvet soft furnishings and marble floors with gilding here and there. It was certainly opulent but, given that the rooms were around ¥30,000 a night, you would expect that.

Gosh, I'm beginning to read like Trip Advisor. Perhaps I should allocate marks out of ten.

I couldn't get over the fact that the hotel was pretty much filled to capacity during the conference but strangely enough, there never seemed to be anybody around. You could walk the corridors, as I did each insomniac night, and meet no one. The bar closed a little after 9 PM, shortly followed by room service. Surely the purpose of room service is to cater for the night owls, suddenly peckish at 2 AM? In the end, I was forced to use a vending machine not far from my room, surviving the nights on a diet of rice crackers and cans of Asahi Japanese lager. Or iced sweet milky coffee on one occasion when I pressed the wrong button on the vending machine.

The sound insulation between rooms was excellent. A good thing except that it contributed inevitably to the all pervading sense of isolation. I couldn't place the source of my unease but something about the hotel also gave me a strong sense of déjà vu.

It was a couple of days before I could place it. Resting between scientific sessions and daydreaming on one of the chairs near reception, an American scientist I knew caught my eye. He looked straight at me, raised his index finger, bending it in time with his words and croaked "red rum, red rum". Instantly I knew where I had seen this hotel before.

It was The Overlook. The famous hotel centrepiece in Stanley Kubrick's film. The colour scheme was the same, the decor similar, even the sense of out of season isolation. I wasn't attending a conference at all – I was an extra in "The Shining". I half expected the elevator doors to open with a rush of blood. Or to see Danny on one of my nightly walkabouts, frantically pedalling along the corridors on his tricycle.

My American scientist friend and I share the in joke. Every time we walked past each other in the lobby, we raised our fingers and croaked "red rum".

"Sorry, bar closed" said a confused Japanese, presumably thinking we were thirsty. I wasn't. I was on my way to pick up a friend for a bit of sightseeing. I knocked on the door.

"Who is it?" I heard.

"Heeeeeeeeeeeeeeere's Jonny".

Sorry. Couldn't resist.

THE JAPANESE TOILET

Forget Pokémon. Forget that long parade of Japanese cars and motorbikes that clog our roads. Forget the Walkman and the Game Boy. No, the greatest contribution of Japanese inventiveness to our Western culture lies not in the automotive or entertainment industry but, you may be relieved to hear, in sanitation. I apologise in advance for any puns – please be reassured that their use is entirely accidental. Or does that make it worse?

In Britain the toilet is a source of embarrassment except obviously to some of the more robust Scouser comedians (and I use the word comedian in its broadest possible understanding). The toilet is an inaccessible little room under the stairs or tucked away somewhere else in the house. Often the room is barely big enough to... well, you know. Not so much a public convenience as a private inconvenience. Blushed cheeks all round.

But, if that seems bizarre, we should not forget that, in many working-class terrace houses, the toilet was not even in the building. It often occupied a small shed appended to the back of the house or even further down the garden. In extreme weather, it would freeze, thus rendering it entirely useless. In those days before global warming, when snow and ice held dominion in winter, entire generations must have gone from December to February without a bowel movement. Not one poo north of Sheffield. It doesn't bear thinking about.

But then the Victorians never actually went to the toilet. Or so we would be led to believe. Presumably. Victorian men were happier visiting prostitutes than reflecting on their own bowel habits. But then Victorian taboos have never made sense.

Japan's different. Anyone who has seen a Japanese game show will be aware that the Japanese capacity for scatological humour is almost infinite. There is nothing about bodily functions that is off-limits it appears. The toilet is close to the Japanese heart.

I first encountered the Japanese toilet at my hotel in Kyoto. Lights slowly flashing and a faint gurgling announced its presence. In fact I have never seen so many lights, dials and controls on something designed to dispose of poo. Well, let's not be shy – that is exactly what it does.

But in Japan, the emphasis is on detail and harmonious interactions. The tea ceremony is a case in point. If you want to watch perfection of movement and attention to detail, then the tea ceremony is its very apotheosis. If on the other hand you are parched, spitting feathers, in search of a cuppa, you have come to the wrong place.

The British toilet is a place to carry out a single function (or two I suppose if you count reading the newspaper). The Japanese toilet is a celebration of imperialist Japanese technological supremacy.

Let's begin at the beginning. Sitting on the British toilet, especially in winter months, elicits a sharp intake of breath and the very real possibility of frostbite. Frostbite in places you don't want frostbite. That very moment of interaction between bottom and Melamine speaks volumes. Specifically it says "get on with it lad". It does not encourage you to linger.

Eight thousand miles away, in Kyoto, Osaka, Nagasaki and Tokyo, the toilet speaks a different language. From the moment you position your posterior on its warm surface (yes I did say warm), it invites you to take your time, to reflect on life. This is a Zen experience. And an adjustable dial allows you to adjust the temperature – from tepid disinterest through to a setting that looked suspiciously like a waffle iron. I left all controls be for fear of being branded.

The Japanese toilet does not stop at the temperature of the seat. No sirree. The toilet is infinitely adjustable. In a small panel on the wall are more ways of modifying your toilet experience than a 1970s hifi graphic equaliser. Behind a mass of – to me at least – incomprehensible hieroglyphs, resided the kind of technology associated with a NASA rocket launch. You could alter the flow of water, its direction, temperature, dispersal and, for all I know, political preferences.

I left all the controls at default. I come from Yorkshire and such fancies are beyond my imagination. In Doncaster, we're just happy not to get frostbite. Leave these pampered posteriors to the southerners. That said, one of my female friends emerged from the toilet with the widest smile I have ever seen. Clearly Japanese technology has its place.

The Japanese toileting experience is not limited simply to the engine itself. Even the environment around can be modified. Lighting can create a mood to match the twittering of songbirds that comprises the default sonic panorama. And if birds are not to your taste, how about serene forest noises or the sound of waves crashing on the shore. Maybe music is more your bag. You can poop to Puccini if you so wish. One thing is for sure – you will never think of Nessun Dorma the same way again. Or for those more decisive moments, perhaps the 1812 overture. There is music for every eventuality.

I am all for technology. And I love to think about the kind of scientists whose life's work is embodied in these magnificent toilets. Over the course of a lifetime, we spend on average, somewhere in the region of six months on the toilet. That's long enough to listen to the entire works of Wagner more than seventy times. And if that isn't a persuasive argument for the application of this technology, I don't know what is. Forget all those Sunderland made Toyotas, this is what we will all be buying post Brexit. You read it here first.

DEFINING FRIENDSHIP

What is friendship? For one reason or another I found myself thinking a lot about this over the last few weeks. On the one hand it seems to be an unnecessary question – we all know what friends are, don't we? Perhaps we do. Perhaps we've never questioned what friends are. Perhaps we just meander through life without questioning what it is that has attracted the people around us to us, and us to them. Perhaps like the big issues of life and death, we'd rather not think about it. Perhaps friendship is like gossamer, the act of touching it being enough to break it. Perhaps it is like a spider's web, enticing and bewitching. Perhaps all of these things and none.

When we were children, it was simple. A friend was somebody who would let you ride his new bicycle. Someone who would be goalie, between muddy jumpers, while you played centre forward. Someone who would play Scalextric with you. Someone who would stand up to Robert Jenkinson, the school bully. Those were friends. Always boys. Never girls. Except Agnes Fairweather, in 4B with freckles and a squint who once shared a bag of sherbet lemons with me. The girls in her class would tease her till she cried. We had nothing in common but that bag of sweets. But that was enough for her to be a friend. At the end of term she told me she didn't really like sherbet lemons. Somehow that broke the spell. I was heartbroken.

I grew older, I moved on and so did my friends. We would see each other occasionally and nod in acquaintance. We became gawky and geeky, with septic faces, spotty and oozing, while the girls laughed at our disfigurement. Agnes Fairweather had her squint corrected when she was eleven. By sixteen, she was 5'8" tall, with ruddy golden curls that reached below her waist. A goddess. All the girls who had once made her cry were desperate to be her friend, to be cool. She didn't remember me.

I moved school, to another town. My friends changed. Old friends from my old school soon stopped writing. Although we had been friends, when it came to the formal act of writing, we did not know what to say. We had never had to think about it. Friends were just the people around you. And there were different people around me now. I never questioned their friendship, never assumed its permanence. Friendship was just there. I had friends and best friends. We all did. It was nothing more complex. And as my old friends had fallen away, often so would these in turn, as I moved on to university.

I suppose the point I'm making is that friendship is very difficult to define or to characterise. The mere act of analysis or definition seems to strain it. The Tao Te Ching, the "Bible" of Taoists, begins with the words "The Tao that can be told is not the true Tao". It makes the same point. In seeking to define it, we somehow find ourselves further away from definition.

The Oxford English Dictionary offers a range of definitions but principally "one joined to another in mutual benevolence and intimacy". It defines friendship as "friendly feeling or disposition felt or shown". Well that's not very helpful. When a linguistic reference of the power of the OED draws a blank, where are we to go?

Perhaps friends and friendship are beyond definition. Or maybe we need to define the word ourselves. I was given an interesting definition today. "A friend is someone you would invite to stay in your home". I can see how that would work, as a practical code. Nobody would ever invite enemies into their home. But perhaps they would ask acquaintances. Where do acquaintances end and friends begin?

There is a school of thought that suggests a key component of friendship is equality – similar give and take between two friends. Again I can see the logic. With such an equal balance, it should be plain sailing. But that doesn't take perception into account. My father-in-law once said to me that the key to a successful relationship lay in the fact that each friend felt the balance to be 90% to 10% receiving and giving. And each perceived themselves as the receiver.

I can see that. I can see how it would work. But I also see an element of exploitation. If you believe yourself to be the major receiver, would you not feel compelled to redress that balance? Would you not feel that you owed more? I once found myself in exactly that position, took no action to redress, and paid a very high price. I deserved it.

But that's not to say that friendship is always equal. There will be times when one friend needs another more than that balance strictly allows. Maybe it's through illness. Fears perhaps. Maybe it's just insecurity. But for whatever the reason, friendships will always be dynamic. There will be times when one shoulders the burden for both. It is a measure of the depth of friendship.

And depth is critical, I think. True friendship is only really discovered in times of hardship. We can all be friends with each other when times are good. When times are hard, it's a different matter. How many of our friends tell us that they are "always there for us" yet are absent when needed most? We make promises we cannot keep. Or more accurately, promises we do not anticipate having to keep.

I had a time, some years ago, when I found myself literally crying on a friend's shoulder. It was in a pub, as embarrassingly open a place as it could be. I had lost someone special and the circumstances had got to me. Eventually I regained my composure and looked up. My friend was looking at her watch.

I've taken a long way to get to this point. In essence I admit that I don't really have a clear definition of friends or friendship. But I think it has something to do with need. Each person fulfils not a wish but a need in the other. I may return to this topic when I have thought more. The answer is out there.

I think sherbet lemons may be the key.

TALKING AND LISTENING IN MEDICINE

We hear a lot spoken these days about the importance of patient involvement in clinical research. Not simply in the sense of being subjects in clinical trials, important though that is, but in the sense of providing intellectual input into study designs and protocols in the form of patient experience. To people like myself with a chronic neurodegenerative (and let's not forget incurable) illness, this makes absolute sense. After all, who knows better the needs of patients than patients themselves? Certainly our consultants will know more about the neuropathology and biochemistry of the condition (well let's hope so). But that doesn't necessarily translate into an in-depth visceral grasp of the impact of the illness. That can only be truly understood from within. As a friend of mine once memorably said "if you haven't got it, you don't get it". And ultimately of course, she is right. No hours in lecture theatres, clinics or case studies can truly help the doctor to understand the true patient experience. In essence, our physicians know what Parkinson's looks like. We, the patients, know what it feels like.

Bridging that gap, between patient experience and physician record, is a charge laid equally at the feet of both communities. Communicating the inner experience more effectively is the task of the patient community. Finding ways to open windows into that experience is the ultimate challenge for medicine in general. It is at the heart of diagnostics after all. These are the challenges that we must address because they are at the centre of understanding the condition. Either alone is the sound of one hand clapping.

But surely we know enough about the patient experience to understand the condition? Clinicians are surely well enough trained to ask the right questions? Not necessarily. Let me give you an example.

I was at a conference of neurologists (yes I know I'm not a neurologist but I am quite persuasive as a gatecrasher). One of their number was presenting a case of a young patient he had observed in clinic and at the patient's home.

Amongst other things he observed that the patient had a strong unilateral dystonia in his hand when lifting a wine glass to his lips. He noted it down as a symptom but did not investigate further. Had he done so, or even spoken to the patient about it, he would have learnt that this was not a unilateral dystonia (i.e. an uncontrollable cramping of the muscles) but a voluntary and conscious effort by the patient to tighten his grip on the wineglass in an effort to reduce the tremor in that hand. The patients knew what it felt like, the physician knew what it looked like.

Most patients respond to that particular story by assuming that the physician was at fault, that he should have asked the patient about what he was observing. But let's be absolutely clear. The blame for this misinterpretation falls on both parties. Certainly the physician should have asked about the symptom he was observing. But the patient should also perhaps have volunteered an explanation of the action. Or perhaps asked the physician what he was writing down.

This is perhaps a trivial example but it's indicative of a wider malaise – assumption. Physicians assuming something is one thing without checking for others. Or, because patients are equally culpable, assuming that the physician will ask all the right questions without the patient needing to volunteer information.

At the World Parkinson Congress in Montréal, six years ago I overheard two clinicians speaking to each other.

"There are a lot of patients of this conference".

"Yes, but I think it's still worth attending".

This is the communication gap we have to bridge. And this will only be done in an atmosphere of equality. Patients – it's time to talk more to the doctors. Doctors, it's time to shed those white coats and ask more questions.

The old patrician model of medicine is a dinosaur. And we all know what happened to them.

A RATHER DIFFICULT CHILD

My mother, God rest her soul, never tired of telling me, albeit often with a laugh, that I was a "rather difficult child". Always highly strung and prone to exaggeration, this was a statement born out of weariness. No longer finding an appropriate hyperbole, she eventually condensed all of my misdemeanours into three words – rather difficult child.

This was borne out by my school reports. "Wilful" was a common theme. "Disruptive when bored" was another recurrent leitmotif. Most of my school marks were As, often qualified with a minus symbol, the inference being that I knew my stuff but was lazy. In my school's eyes, a B+ was a more honourable grade, suggesting an honest hard-working plodder. Good Protestant middle-class values. A grades were an admission of excellence and, unless you were also the captain of the cricket team or your father has just refurbished the library, were always qualified with a minus. There always had to be some area of deficiency in your school report, perceived or real, with which your parents could browbeat you. It was just the rules.

I sailed through prep school, always finishing either first or second in the end of term exams. Generally second if I'm honest behind Stephen Halliday whose grandfather had once been a cabinet minister and could recite huge chunks of Shakespeare. This was Yorkshire in the 1960s. You were hard pressed to find anyone who could spell Shakespeare, let alone recite it.

If I was third in the end of term exams, my mother would greet my father from work, wringing her hands and hopping from foot to foot in a maelstrom of anxiety. Jonathan had come third – the world was evidently about to end. At the very least, she would be unable to go to her regular hairdressers in Bawtry except in dark glasses. The shame of it.

"Speak to him" she would say to my father. My father, engrossed in the Telegraph and a pre-prandial glass of Bristol Dry, would briefly lower the newspaper, raise an eyebrow and, unnoticed by my mother, wink. Once.

In contrast to my mother, my father had a yen for the dramatic understatement. Like emperor Hirohito whose surrender speech concluded that "the war situation has developed not entirely to Japan's advantage", so it was with my father. A Cambridge scholar, brilliant diagnostician and creative thinker, he took such qualities as granted in his offspring. It worried him not one jot that his eldest had come third in one exam.

The fact was I was brilliant at school. Nobody ever told me otherwise. I was everything you would expect from such intellectual genes. Arrogant. Conceited. In spades.

It was not always so, my mother said, a year or so before she died. "Remember kindergarten".

I did indeed remember kindergarten. I stood out. "I was in Miss Poskitt's reading class" I said "she called us special".

There was a pause before she spoke. "I never had the heart to tell you"

I looked up. "Tell me what?"

"Did you never wonder why it was called the special reading class?"

WITNESSING HISTORY

I remember being suddenly awake, with my father at the foot of my bed.

"Jonathan" he said, almost in a whisper "come and watch this".

It was the middle of the night, way outside any normal television hours. Broadcasting in those days finished before midnight with the national anthem. There was certainly nothing to see after that.

But tonight was different. From my end of the corridor, I had been vaguely aware of my father's alarm clock some quarter of an hour earlier, like a distant fire bell in the night. But somehow my mind, weaving in and out of consciousness, had failed to register it as significant.

"Wake up" my father said, a little louder, shaking me gently.

"I am awake" I croaked from dry lips and throat, blinking at my alarm clock with its unsettlingly bright radioluminescent dial and hands. Three o'clock it informed me through its usual hailstorm of beta particles. Can't believe these things were ever legal. But that's an aside.

"What's happening?" I asked.

"Come with me" he said "and don't wake your sister".

I pulled on my dressing gown and slippers.

Mother was in the living room and the television was on. There was a smell of coffee, brewed for herself and my father. Not for me – coffee was not for eleven-year-olds.

Nothing made sense.

"We're going to watch history" said my father. My heart sank – history during school hours was one thing, its intrusion into my night's sleep quite another.

"One of the American astronauts is going to walk on the moon" he explained and, as he did so, the television picture shifted to mission control in Houston and the banks of computers that had taken the Apollo program to this point. Some seven years after President Kennedy had announced his country's intentions to go to the moon, history was fulfilling itself. In a matter of minutes, the drama was to unfold.

I yawned,

"Turn it up" my mother said "I can't hear anything".

There were no remote controls for televisions in those days. If you wanted to adjust the volume, you got up and adjusted the control on the box itself, sat down, realised you were on the wrong channel, got up, pressed the right channel, sat down again, noticed that the screen was too dark, got up again, adjusted the brightness, sat down again, had a shower. It was an aerobic workout. And, back then, there were only three channels to choose from.

In the Stamford household, we didn't even have colour. My father thought it a rather vulgar modern affectation, an insult if you will to the ghost of John Logie Baird. I still maintain that we were the last house in Doncaster to have colour television. Sometime in the late seventies.

"Why can't we have a colour television?" My siblings and I repeatedly requested.

"Reg and Nina have it" he said "it just wouldn't do".

Reg was the bookie who lived next door. He had a colour television, the colour permanently turned up to maximum. As an accurate representation of the outside world it was a lamentable failure. But as an electrical embodiment of a Grateful Dead concert, it could not be faulted. News broadcasts were delivered by orange aliens. The horseracing became a blur of psychedelia. If Nina had been popping amphetamines, she could not have done better.

Either way, we were left to watch the complexities of snooker without benefit of colour. No one should be forced to do that. Or watch snooker at all for that matter.

But back to the moon landing.

Just as my father finally adjusted the television to his satisfaction, the hatch door on the lunar module opened. It was just about possible to make out this detail. Pictures were terribly faint, almost lacking in contrast, focus or anything approximating to clarity. I remembered grumbling to this effect and being reminded that the pictures were coming from the moon. That we should have any pictures at all was a testament to that huge bank of computers at mission control. Incredible really when you think that there is more processing power in a smart phone than there was at mission control in the 1960s. Such is the mark of progress.

There was a lot of beeping. Slowly at first, and certainly ungainly, Neil Armstrong descended the ladder, rung by rung. His spacesuit was pressurised, my father told me. If he got a hole in the spacesuit, he would explode. I remember thinking that would be pretty cool.

Then, in an instant, it was done. He leapt, as though in slow motion, from the last rung onto the surface of the moon. There was a pause. Even across the distance to the moon, you could hear the silence.

"That's one small step for man, one giant leap for mankind". Or, as it was on our television,"SSShhat'sss ssswwwone sssscchmall sssshhhhtep for man, sssshhhhhone giant leap-p-p-p-p for ssshhmankind". Equally memorable, in its own way, I'm sure you'll agree.

And that was it. That was what it felt like to witness history.

The Apollo XI moon landing was the apotheosis of all the optimism of the 1960s, a decade when anything seemed possible. It was the fulfilment of the dream of a great American president. It was a time when America truly was great. America was led by strong thinking men who addressed their country's inequalities head on, not cowering behind a wall. A country built on post-war immigration. A country built on hope, not fear and led by men who knew the difference between great oratory and infantile rhetoric, between promise and puffed up prejudice. As different as chalk and cheese.

Fifty years ago. It seems like a lifetime.

HAIRCUT

Yes I know I've said this before but in case anybody missed it or is new to my writing, I grew up in Yorkshire in the 1960s and 1970s. Specifically I grew up in Doncaster, the erstwhile self appointed capital city – never mind that it was only a town – of The Socialist Republic of South Yorkshire (Arthur Scargill, president). I say South Yorkshire but of course what I really mean is the West Riding, the administrative division of the county from the late 1880s until 1974.

Doncaster was surrounded by coal mines. Askern Main, Brodsworth, Bullcroft, Frickley, Goldthorpe, Hatfield, Hickleton Main, Rossington, Thorne and Upton. Even reading those names now brings a tear to my eye. They are Yorkshire poetry.

And that's just the names of the collieries – the villages themselves are even more romantic. Woodlands, Carcroft, South Elmsall, Goldthorpe, Stainforth, Thurnscoe, Wath on Dearne, Armthorpe, Moorends and Edlington. Names to conjure with. Names synonymous with hard graft. Gritty names redolent of honest endeavour. For nearly a century, these great pits dug the black gold that fuelled British industry.

Sure there were pits in other parts of the British Isles. And they too exemplified some of the same values. But these pits were in Yorkshire. God's country, as any Yorkshireman will tell you. As Yorkshire as Eccles cakes, Yorkshire pudding, whippets and Wensleydale. Yes I know that Eccles cakes technically come from Lancashire but we all know that is simply an accident of geography. Besides, my aunt Paddy who made the best Eccles cakes known to humanity came from Yorkshire. Or lived there at some point. Eccles cakes are therefore a Yorkshire food.

But nothing encapsulated the uncompromising nature of the Yorkshireman at large than the Demon Barber of East Laith Gate, Ray Fowkes*.

Ray was a barber. Not a hairdresser. He was quite clear on that point. He knew only one hairstyle – the short back and sides. To call it a hairstyle was in itself a misnomer. And to be honest, I never heard Ray use the word style in the entire decade he cut my hair. It was a haircut and if you didn't know the difference, Ray was more than happy to explain. At length. Often his explanations included his thoughts on immigration, teenagers, pop music and the enduring value of national service.

I think it would also be fair to say that Ray wasn't much of a listener.

Strangely, although he knew only one haircut, he executed it with precision. It took time. Sometimes quite a lot of time, depending on where Ray was in his latest soliloquy about Enoch Powell or nationalisation. Or his most popular mantra on the relationship between long hair and mental dissipation. I still remember the day when a young man – evidently not a local – explained to Ray that he was in a hurry, and would it be possible to have a quick haircut.

The entire salon (Ray never called it that) fell silent.

"You can have a quick haircut or you can have a good haircut" said Ray calmly.

It was like a Bateman cartoon. You could have heard a pin drop. I remember wincing as the man expressed his view that speed trumped accuracy. Ray nodded in acknowledgement. I remember biting my tongue three minutes later when Ray had finished his interpretation of the man's instructions. It was indeed a quick haircut. Technically at least, in the sense that he had cut the hair. But it was probably closer to a shearing. There was something vaguely agricultural about the tufts and patches. You half expected it to reveal a branding mark.

"Did you have to do that, Ray?" asked one of the regulars.

"Aye" said Ray "lad's got to learn".

The lad did indeed learn – principally he learnt to go to a different barber. As did another who turned up in Ray's shop with a copy of a David Bowie album. For maybe a minute, the young man earnestly pointed out the key features of the Ziggy Stardust hairstyle. Ray nodded at each juncture. Half an hour later, Ray was finished. The young man was plethoric.

.

"Ziggy Stardust's hair doesn't look like this"

"It would if he came in here" said Ray. "That'll be eleven and six".

*Not his real name...

GRAND RAPIDS DIARY 1: LAND OF THE FREE

Yes, it's time for the annual tickling of the Parkinson's brain cells that we call RALLYING TO THE CHALLENGE. It's a meeting where, although primarily for patients, is amply supported by our clinical colleagues. But for me, RALLYING 2019 has a certain edge to it, for two reasons. Firstly, the focus of the meeting is genetics. And secondly, I have been charged with responsibility for putting together the program.

Now on the face of it that looks like a hospital pass for the simple reason that the sum of my knowledge on genetics in Parkinson's would, at best, cover perhaps a couple of postage stamps (definitive not commemorative). Fortunately, I really have been able to stand on the shoulders of giants and have been overwhelmed by the quality of the scientists who are going to speak at the meeting. So it looks as though I will be able to carry this off, appear at least halfway knowledgeable, and deliver an excellent meeting. Of course, declaring my own intellectual limitations in genetics is probably foolish. There are plenty of people out there ready to trip me up with their triple codon exon intron sequence phosphorylation base deletion phase shift antisense misreads. But I have my fake coronary routine ready for that one, should I need it.

The conference week has not however started well. And I'm sorry, but I am going to name and shame.

We left Heathrow on time, were ahead of schedule until we got to near Chicago. Then we circled for an eternity before finally landing. Even then our parking space was taken and we had to wait a further thirty minutes on the tarmac before being allowed to disembark. Then of course there is American immigration. Don't get me started. Ludicrously complex, counterintuitive, contradictory and.. Oh something else beginning with C. It was not how the day was meant to go.

We had breakfast at Heathrow, supper in Grand Rapids only to find our luggage still in Chicago. Or actually not even that. We last waved our luggage a cheery goodbye in Chicago, placing them in the hands of the most lackadaisical and disinterested baggage handlers you could easily imagine. Don't get me wrong – a suitcase is a suitcase not a Leonardo fresco, or a rare rainforest orchid. I don't expect it to be treated with reverence but throwing the thing onto the conveyor belt in front of its owner and in such a way that it's safety straps were partially dislodged is not the action of a man comfortable in his employment. I don't imagine for one second he wakes up each morning, aching to be of service to his fellow man. But it's sad to see such open contempt. And of course I don't know the back story. Maybe he argued with his wife. Maybe the kids are failing in school. There are a thousand reasons why my suitcase instead of sitting in my hotel room is currently languishing, and possibly damaged, somewhere between Chicago and Grand Rapids.

But of course it does have larger consequences. As always I deliberately try to travel light. My cabin baggage consists of a spare day of medication (in case my suitcase gets lost – you see how this is shaping up), my phone, laptop and sunglasses. The chargers for the phone and laptop are in the suitcase. As is my wash bag, cameras, tripods, gimbals, microphones, several days clothing and last, but no way least, my medication.

That's right – the medication I have to take every three to four hours, just to stay alive and functional. I don't mind losing all the electronic gear (actually I mind a hell of a lot) but the prospect of being without medication fills me with fear. To put it in a nutshell, I have until the end of today to resolve this and to ensure that I have enough medication for the rest of the week.

Of course, if you have Parkinson's and are short of medication, it does help if you are attending the meeting of some hundred other people with Parkinson's. At least someone there is going to be able to lend you the necessary pills.

I suppose I should put the usual disclaimer here – people with Parkinson's should not share their medication with others, or words to that effect. And, yes, I do endorse that principle. Just not today.

I'll keep you all updated.

GRAND RAPIDS DIARY 2: LUGGAGE

Where were we? Oh yes, I remember – in Grand Rapids without our luggage.

I'm pleased to say that the luggage did eventually show up, late morning and somewhat battered. But to be honest my sense of relief simply having it in my possession again more than counterbalanced its state. In any case, by mid-morning I had received several offers of clothing from Australia to Canada and all points in between. I was touched by their generosity. Although it is equally possible they were aware of the environmental consequences of me wearing the same underpants for 48 consecutive hours. Even Greta Thunberg was alerted.

Okay, I lied about the last bit. But there is no doubt that my undergarments were beginning to acquire a personality all of their own. And not a particularly nice one either. Whereas the wearer is gentle soft soap, liberal socialist, the undergarments are a more Tommy

Robinson/Brexit/ rent a thug sort of persona. You know the type. Definitely not the kind of underpants you want lurking at the back of your knicker drawer.

More or less everything had been put on hold while we waited for our luggage to finally appear. It's hard to plan a set of slides and handouts whilst simultaneously wondering whether the time would be better spent in buying clothes. If the luggage had not arrived before 5 PM for instance, I would have been faced with the very real possibility of having to deliver a scientific presentation wearing only my underpants.

The scientific community deserves better.

Anyway, all's well that ends well. The rather overassertive underpants have been neutralised and I stand resplendent and fragrant in my purple boxers. Ladies, form an orderly queue.

Yesterday (Monday) was fundamentally a rest day before the onslaught. Gradually people arrived throughout the afternoon. The proceedings really began with the reception at the Van Andel Institute in the early evening. A bottle of beer and a very generous glass of Glenlivet more or less put paid to me for the evening. I hit the wall so to speak, ducked out of the organised dinner and started work on my presentation. Well that was the plan. Somehow it ended by having dinner with Heather and meeting up later with Gaynor for a nightcap.

And the presentation? Oh yes, I remember. Specifically I remembered at 4 AM. To say that I sprung out of bed would be an overstatement. But certainly the pressing urgency was clear to me. Four hours later, the presentation is finished, meetings have been arranged with other speakers and the documentation for the meeting was finally in place. We have a day left for run-through, so it's looking good. A couple of quick notes to speakers and helpers and all is done.

We are ready. Bring it on.

GRAND RAPIDS DIARY 3: NIGHTTIME, DAYTIME

You forget how tough these conference trips can be. Late nights and early mornings are okay when you're young. But as the years mount up, each exacts a heavier toll.

As you know, well those of you that read my random scrawls know, sleep is everything for me. If I get a good night's sleep, and I'm talking somewhere around four hours, then everything is pretty much okay with the world. I can function. I know what time of day it is. I can even make sensible decisions. All the appearances of a reasonably high functioning individual. I say appearances because obviously it's partly illusory.

Five hours is even more valuable and six – well I hardly know what to say it's so rare. I can't remember when I last slept long enough to be woken by my alarm clock. Not entirely sure it was even in this millennium.

The point I'm making, albeit somewhat circuitously, is that my sleep is fragile. Like a cobweb or a chrysalis. Almost infinitely delicate and easily disrupted.

So when you throw in a five-hour phase shift by moving from the UK to Michigan, it hits me hard. Much harder than most people I fear. For most of us, and by "most" I mean those upon whom the gods of sleep see fit to bestow seven or eight hours, jetlag is perhaps a gentle punch to the midriff. For me, it is the equivalent of a rapid left-right combo to the head topped off with a haymaker to the groin. It fells me.

I'm awake when others are not. I have padded the hotel's corridors at three in the morning in my slippers like some homeless spectre, met janitors clocking on for the morning shift, watched steam rising from the storm drains, spoken to the street sweepers, clattering their brushes on the dustbins. I have burnt the midnight oil, the 1 AM oil and so on. Oils for every hour. The ghosts of Saturday night, as Tom Waits would say.

It almost sounds romantic, the city in another light. Like Monet's pictures of Rouen cathedral.

But it isn't. It's desperate stuff. When I'm awake others are not. But when others wake and function, I am distant, withdrawn, unfocused. Grumpy even.

People think these trips are easy, a jolly even. They're not. It's hard work with sleep, harder still without.

I've probably laboured the point. I'm not the only one who is tired. But we have got through to the end of the day. On the whole the presenters stuck to their allotted times, didn't veer too far from their briefs and entertained questions in the spirit they were couched. And there were some excellent talks. I often get the feeling that I understand genetics less well than my fellow man. These conferences turn those nebulous inclinations into rigid certainty.

But you are a scientist, I hear you say. Well yes, certainly I used to be. But genetics really is big science. And it's been a long time since I paddled in those pools. It's fun listening to the debate for two reasons – it's often at a very high level. And it's conducted largely by patients.

Time and technology moves on. I'm turning rapidly into the kind of person who can't operate a video recorder remote control. Stuff my kids take for granted leaves me floundering. And when it comes to time and technology, genetics and the technology genes has outstripped practically anybody's capacity to understand it.

And if I can't understand it on four hours sleep, I'm never going to make it on two.

Better leave it to the youngsters, I say.

I'll just catch up on my sleep.

GRAND RAPIDS DIARY 4: WE DO OUR BEST

It's the same every year. We put everything we have into this meeting – RALLYING TO THE CHALLENGE – each year. We burn the midnight oil. We meet and greet, we talk, we listen. And while the meeting goes on, a dozen little things take place to ensure everything runs smoothly. For every speaker who speaks, there are AV technicians who make sure the sound is just right. There are runners making sure that the next speaker is ready and prepared, on standby. There are people talking to advocates and inviting them to question. There are last-minute changes to the running order because maybe one speaker is stuck in Detroit because their connecting flight was cancelled. There are audio and video links to be urgently prepared so that they can still speak seamlessly at the meeting. Then there is timing – it's no use calling the tea break before your support staff have the cookies ready! So somebody has to keep an overall view and liaise with the session chairman that they are running too fast or too slow.

All this happens without the audience participants being aware. While they see the swan gliding across the lake, they do not see the frantic paddling beneath the waves to maintain that image. Next time you are a conference, bear that in mind when you complain that the butter pecan cookies aren't as sweet as you like. Or the vegetarian options are less exciting than you would hope.

We do our best. Dear God we do our best. And, often we feel appreciated. You tell us that you like this and that's why we keep coming back each year, from our base in London to be with you in Grand Rapids. Familiar faces, new faces. Engaged faces, interested faces. That's what makes it worthwhile for us.

One delegate at the meeting came up to me to speak. It was his first attendance at these meetings. "You know" he said "this meeting is really intense".

You think so?

To be honest, we love it. And we love the fact that you welcome us back each year. But by the end of the meeting, we have nothing left to give. The CPT team stands by the entrance to the Van Andel Institute, waiting for our taxis to take us to the airport this. We have twelve hours of flying ahead of us. Grand Rapids to Chicago then on through the night back to Heathrow. We hardly make conversation. We are that tired. Spent forces. Our minds are whirring with everything we have seen and done over the last couple of days. We turn things over in our minds. Should I have said that? Did they really mean what it sounded like? If they really meant that, why did they not say so? A thousand thoughts and reflections. Each of us is absorbed in our own small world of memories. Some sit in chairs sleeping. Others unpack and repack their suitcases.

There are a hundred goodbyes. Emotional goodbyes. These are, after all, our friends. To say goodbye to one is a wrench. To bid a hundred farewell is the sweetest of sorrows. And you wonder why there are tears shed?

The taxis arrive. We pile in. Thirty silence filled minutes later we are at Grand Rapids. The bags are checked and our long march home begins. Five hours later, we settle into our seats on the plane from Chicago. I have held it together till now when, for a thousand little reasons, I am overwhelmed by a tidal wave of emotion. I sit in my seat, bent forward and sob uncontrollably. It's not a pretty sight. And I don't even know all the reasons. Emotional wreckage. I am absorbed in myself. It's almost as though people pass by in silent slow motion. I feel dissociated.

We get to Heathrow eight hours later. Some sleep, some do not. There are more tearful farewells by the luggage reclaim. It's the weekend. We are drained. We gave it everything.

And you know what – we will do it all again next year.

LONG DISTANCE FLYING WITH PARKINSON'S

Travelling is arduous even for the best prepared, keeping tabs on passport, currency, boarding cards and so on. Flying is, of all transports, perhaps the least passenger friendly. There is no time to make oneself comfortable while being herded like cattle.

And when you have Parkinson's, those difficulties are compounded by shakes, freezing, unpredictable dyskinetic and dystonic movements. The list is endless and it makes for a tough experience even on short flights. Long haul however is my ultimate nightmare.

Let me give you a brief rundown.

Firstly, my tremors are far worse on travel days, probably because of the anticipated stress. I am sweaty, uncomfortable and shaking like a leaf often before I have even boarded an aeroplane.

Secondly, I have restless legs syndrome, a particularly uncomfortable component of Parkinson's and one which means I'm constantly having to get up from my seat to walk about until the symptoms abate. And if you've never had RLS, let me assure you it is far worse than its rather genteel name might suggest. Imagine your legs writhing internally like cans of worms. Get the picture? Not at all nice.

Thirdly, I have rem sleep behavioural disorder (RBD). This means that I hardly ever sleep on aeroplanes. I can't because I sometimes thrash about and act out my dreams. Alone in my bed at home this is not a problem. The only person who gets hurt is me, falling out of the bed or punching a hard wall. But on an aeroplane, this is catastrophic.

Fourthly, I need the toilet frequently. Bladder urgency, even frank incontinence, is one of the less widely touted symptoms in the smorgasbord of misery on offer to the average Parkie in the street. But you simply cannot be caught short on an aeroplane. So I don't drink much. Even so, if the toilets are occupied I become anxious, which makes the tremors worse and exacerbates my other symptoms.

Fifthly, and this is a relatively recent phenomenon, I freeze. And always at the most inconvenient times. My neighbour will perhaps ask me to get up so that they can go to the toilet. My inability to move is taken as rudeness and my anxiety levels and attendant tremors go through the roof.

Sixthly, and especially when I'm nervous, my voice becomes quiet. I have to repeat things, often several times. On an aeroplane, where noise levels are high, I simply can't get enough voice out.

All these facets make travelling by air something to be dreaded rather than anticipated.

But what is the solution?

Well, it's not foolproof but I do make certain preparations where possible. Many are obvious – I try to use the toilet before I get on the plane, I don't drink alcohol, I try to make sure that my tablets are taken at such a time that their peak effect will be whilst on the plane. And so on. Lots of tiny adjustments. But more important than any of these, I try to travel with a "buddy". This should be somebody who knows I have Parkinson's and can make appropriate accommodation to my problems. In a perfect world, it's a good friend who knows me so well that they can anticipate problems before they occur rather than after. This often means they are a Parkie themselves. I have a tiny handful of such friends and knowing that I'm travelling with them beside me makes all the difference. I say beside me because that's what I mean. They can only offer help if adjacent, not five seats away or beyond. And there is a double benefit because, in turn, I look out for them.

At least this way I have a fighting chance of making it to the other end without mishap.

But when it goes wrong, it goes royally wrong.

I was on a recent flight to Heathrow from Chicago. An overnight flight, so already a potentially tough situation. But I was happy and confident because I had my "buddy" in place, with the ticket immediately to my right. Better still it was a fellow Parkie and a good friend. Everything was set. I couldn't guarantee how the flight would go but I was at least ready. I was in safe hands. I had stacked the odds in my favour.

Then, out of the blue, and for reasons that are unimportant any more, I was bumped to a different seat many rows away with more legroom. I remember shaking the moment I was told. I was angry and frightened. Suddenly my safety net had gone. I didn't know who I would be seated with.

It turned out to be two total strangers. The one immediately to my right was about as unfriendly as you can imagine. A dark-haired grumpy troll of a woman, she didn't even speak one word to me during the flight. She pointed to the seat beside me, no "excuse me", "please" or "thank you". I got up slowly as I do and although she did not tap her fingers, her impatience was clear. She sat down, wedged her bag under the seat in front, fastened her seatbelt and sighed. I tried to make eye contact but she looked away. I felt like a leper.

My tremors went out of control at this point. I sensed her slide over slightly away from me, repulsed. Evidently she thought I was a pervert. She put on her headphones as I tried to explain.

The journey was every bit the nightmare I expected. I could not get to sleep, not that I dared to anyway. I walked up and down the aisle until the stewardess asked me what I was doing and suggested I went back to my seat. My legs were writhing like mad with RLS. Can you imagine an itch you cannot scratch? Over your entire legs? No, I don't suppose you can.

I got up again, hoping my buddy might be awake to sense the distress I was in. Alas not. I briefly thought of waking them. There was nobody I could talk to. I'm ashamed to say, but it's a measure of my distress, that I went back to my seat and wept. Trollwoman was unmoved.

But the worst of it all was breakfast. As dawn broke, the cabin staff served breakfast. My tablets were in my satchel but by now I was so late taking them that I was frozen. I couldn't even peel the lid off the yoghurt pot. And when a cup of hot coffee was placed on a tray, I panicked. The tremors were immediately out of control and I knew I was going to knock the drink and food onto the floor. Trollwoman continued watching the film. Eventually I caught the eye of a friend who, after a bit of gesturing and mouthing "help!" (I couldn't speak) came to my aid. And that's pretty much how the flight ended – with me frozen and sobbing.

In short, an absolute nightmare flight. I felt diminished, degraded and humiliated.

They say that typhoons in the South China Sea are triggered by the flapping of a butterfly's wings in Tibet. In other words small actions can have catastrophic consequences. So it was here. I don't know for certain that everything would have been a success in my original seat. But at least I had stacked the odds in my favour.

Those of you reading this who do not have Parkinson's may dismiss what I've written as hysteria or similar. You may wonder what all the fuss is about. What is this RLS and RBD? But those who do have Parkinson's may recognise such situations.

But at least you no longer have to ask me why I hate flying. Now you know.

BETTER PATIENT ADVOCACY
1: STANDING ON THE SHOULDERS OF GIANTS

It is an article of faith in science that discoveries build upon previous knowledge. Sir Isaac Newton, discoverer of gravity inter alia, expressed this better than most when he said "if I have seen further it is only by standing on the shoulders of giants". In essence, whilst downplaying his own contribution, he acknowledged the nature of scientific research, discovery building upon discovery.

This is an important concept and equally valid today. Science does not reach conclusions without building upon solid foundations. Often those foundations are bound together by knowledge from different branches of learning. In some ways even that diversity may contribute to the strength.

In recent years, we have seen a substantial conceptual shift in the role of patients in medical research and discovery. Patients are evolving away from their traditional role as the guinea pigs of the scientists. Increasingly patients are involved in reviewing scientific applications, contributing to the process itself as often as not. There are even whisperings of patient driven and patient initiated research in the near future.

In many respects, these role shifts are the result of many years of patient campaigning and advocacy. We, as patients, take it as our right to be involved at the very core of the research endeavour. In essence we adopt the age-old motto "no decisions about us without us". Like a country's constitution, we hold it as quintessential that we have the right to represent ourselves, that nobody should speak on our behalf. To all intents and purposes, we demand parity with our scientific colleagues.

And herein lies the problem. We are not scientists. We are in essence asking for roles and responsibilities that in many cases we are not ready to implement. Put bluntly, we are often out of our depth.

Don't get me wrong – I believe entirely in the principle of patient involvement. It's not that it is the best way forward so much as the only way forward. Research about patients without patients is an absurdity. Of course patients have to be involved. We just need to find out how.

Let's return to Newton for a moment, his apples and his giants. Or more accurately, their modern counterparts in their laboratories around the world. When writing grant applications or research publications, it is taken as read that the modern research scientist is absolutely up-to-date on the latest research elsewhere, that they can recall their findings and place their own work in its appropriate context. In essence, they know the exact identity and detail of the giants upon whose shoulders they are standing. Research is iterative. It has to be. It builds on previous discovery and projects forward by inspired intuition and happy informed guesswork. Knowledge and understanding of "the literature" is imperative.

Theories created in isolation rarely find favour. Not because they are necessarily inherently flawed but because they fail to acknowledge the importance of previous discovery.

At worst, this amounts to rediscovering the wheel. Not surprisingly, grant awarding authorities take very little interest in work that shows such poor scholarship. Publishers likewise spurn manuscripts that fail to acknowledge the primacy of others' work.

But what, I hear you ask, does this have to do with patients and patient involvement in research? We're not scientists, you say. Certainly, but we aspire to equality in respect and understanding. And although we have those aspirations and seek those roles we are not, as I stated earlier, universally equipped to do so. And I believe the reasons for this are simple. We, as patients and advocates, stand at ground level. We do not stand on the shoulders of giants.

Parkinson's is a cruel mistress. As the years go by post diagnosis, we shift from timid ignorance to vocal experience before gradually disappearing again, raging against the dying of the light. It is one of the most brutal ironies that one reaches the greatest understanding of the condition only as one's ability to communicate that knowledge dwindles to the sound of silence.

This is the problem. These are our giants if we only but recognised them. And, if you will excuse the following excruciating mixing of metaphors, we rediscover the wheel because we do not stand on the shoulders of giants.

Let me personalise this. I have had Parkinson's for around thirteen years. During that time I have witnessed – even been part of – many initiatives aimed at improving quality of life, better understanding the condition and even hastening a cure. Often these initiatives were led by advocates now gone. And as they faded away, so did their ideas.

And every few years, a new generation would appear, full of energy and inspiration, brimming over with "I'm different, I'm going to beat this thing". And as the new generation of leaders emerged, so did their followers. Chat rooms and discussion groups would emerge, with different names but strangely familiar content. Old issues have been recapitulated.

But the one thing signally absent in this process is communication between those dynamic young firebrands and the flickering embers of the old guard. The young were too busy to listen to the old and the old too self absorbed with simply surviving for there to be meaningful exchange between the two.

This is a terrible state of affairs. The older, or perhaps I should simply say more experienced, patients have walked the same roads that the youngsters now tread. If the youngsters looked hard enough, they would see their footprints. They would see where discoveries had been made or ideas refuted.

But this isn't about the inability of the newer generation to listen to their forebears. It is as much an admonishment of the older generation for failing to pass their ideas on to those best equipped to implement them.

This is where we differ substantially from scientists. Whereas their very success depends on their knowledge of what has been done previously in their fields, this is not the case in patient advocacy. The younger generation are not absorbing or even aware of the treasure chest of knowledge to be tapped by conversation with the previous Parkinson's advocacy generation so to speak. And the older generation are failing to show the youngsters where the treasure chests are. This I sincerely believe is essential if we are ever to claim our rightful positions at the tables of research charities, policymakers and discoverers. We have to find ways to build on the experience of previous generations. Only then will we truly be standing on the shoulders of giants.

Further recommended reading:

Jean Burns, "On the shoulders of giants". https://pdblogger.com/2019/08/22/on-the-shoulders-of-giants/

BETTER PATIENT ADVOCACY
2: PASSING THE BATON

Until the 1960s and the advent of L-dopa, the lot of a person diagnosed with Parkinson's disease was a pretty miserable one. Typically six years from diagnosis to death, and six rather unattractive years at that. Progression was largely untouched by the drugs available which, in any case, carried a significant cognitive payload. Six years in which to put your affairs in order, squeeze the last pleasures from one's former life and come to terms with one's imminent extinction. No wonder my mother, who had nursed end-stage Parkinson's patients in the 1940s and 1950s sobbed when I told her of my diagnosis.

But the introduction of L-dopa in the early 1970s changed that picture significantly. And although it did not cure the illness it delayed the deterioration to some extent. A life expectancy of six miserable years became twelve years, eighteen years and beyond. Parkinson's patients who anticipated a brief agonised exit suddenly found themselves with time on their hands. And although that single observation alone is not enough to explain the rise in advocacy over that timeframe, it is most certainly a contributor. It is my firm belief that medical conditions with relatively long post diagnosis lifespans are the breeding grounds for advocacy.

Let me explain. I believe that the best patient advocacy is the product of a relatively long post diagnosis lifespan and a poor and deteriorating quality of life. In essence two interacting facets. A rapidly terminal illness, such as one of the many cancers, affords the sufferer little time to do anything much beyond write their will, say goodbye to relatives and set their house in order. It certainly doesn't allow sufferers to build a useful programme of patient advocacy. Conversely, conditions with a long lifespan but little deterioration, although providing the timeframe necessary for advocacy, to not have the necessary burden of illness.

Both a lengthy period of illness and a significant and increasing burden are necessary. In these conditions (Parkinson's, Multiple Sclerosis, cystic fibrosis for instance), it is not surprising that patient advocacy flourishes. We should be grateful for that. And in many respects we are. Advocacy is our way of drawing people without the condition into our world, of helping people understand what we go

166

through. Because otherwise "if you haven't got it, you don't get it" as a fellow patient once said to me.

Advocacy serves its purpose, if its purpose is considered to be that of raising awareness among the general public, attracting funding and thus increasing the amount of research. Laudable aims but somehow still not a very high bar. For many Parkinson's advocates this is not enough. We expect to be involved in all aspects of the condition from diagnosis, through research and the lived experience, to quality-of-life and end-of-life issues. Nothing falls outside our perceived remit.

And, as outlined in the previous article, therein lies the problem. We have fought and achieved roles in most aspects of research and care. We have learned much along the way and have been inspired by many brave and imaginative people. But where we have failed, and perhaps it is simply a victim of our success to date, is in transferring that knowledge and understanding from one Parkie generation to another. Knowledge accrued by one generation seems somehow to be taken to the grave by that same generation. Either they have failed to communicate their learning to the youngsters with the energy to run with it or, equally likely, the newer generations simply never knew what had already been established. So much is lost, as Rutger Hauer memorably said "like tears in rain".

We have to find a way to value the knowledge of the older generations. We have to find a way of recognising their value as, in essence, the giants upon whose shoulders we stood. I don't believe we are currently doing so. I believe so much of that knowledge is lost. As the condition progresses, it inevitably diminishes our capacity to communicate. Whether we recognise it or not, eventually our diminished powers of communication somehow ossify the knowledge gained over a lifetime with the condition. The older generation watches as the youngsters make the same mistakes, hit the same brick walls.

Each new generation rediscovers past learning, often without realising. They fail to notice the giants watching their every move with interest. And the giants, paralysed with the burden of the years, let each firebrand pass without speaking up. This is a problem. If we do not, collectively and individually, pass on our knowledge as we approach the autumn time, that knowledge and learning will be lost. We need to find a mechanism of ensuring that the baton is passed in a systematic and helpful way. But how?

It would be warming to believe that it can be left to individuals to buddy up with advocates from different generations, mentoring the younger whilst still listening to our forebears. It's an appealing notion but not one that is readily scalable. Relationships like that build organically – they cannot be imposed. Success is predicated on the basis of personal relationships. Ultimately this is insufficient to carry the baton forward in a meaningful way. What is needed are stories - the basis of a collective knowledge.

Let me explain. Many of the more ancient cultures still surviving today owe their culture to the oral tradition. In many cases, written language appeared later. But it was the oral tradition – spoken stories – that was passed from generation to generation.

Before the influx of voluntary (and some less voluntary) immigrants to Australia in the last three hundred years, the country was populated, albeit sparsely, by aborigines for over fifty thousand years. And during that time, the aborigines maintained a lot of their cultural identity through repeated stories of a mythical prehistory. This prehistory, The Dreamtime, was handed down orally from father to son over more than a thousand generations.

In North America, Native Americans lived under the eye of Wakan Tanka and practised animistic rituals to appease their multiple deities. Again these traditions were oral, passed among tribes and down lineages. Nothing was written down in anything we would consider written language.

Even in Europe, with its widespread intermingling of populations through trade and war, there are traditions. We owe our understanding of the great Nordic sagas to books. But it should be remembered that these stories predated their littoral transcription.

The point I'm making is that the oral tradition is a powerful means of communication and information transfer down the generations. The ancient peoples were unencumbered by the need to write down stories. Consequently their oral traditions are all the stronger.

This is telling us something. Something both informational and sociological. We have to be able to transfer knowledge of the elders to the youngsters. I take that as read. We have to find a way of passing the baton that negates the generational mistrust prevalent in modern society. We have to find a way in which the youngsters will wish to hear the wise words of the elders. Not only will this prevent us reinventing the wheel but it will also surely strengthen the sense of community amongst people with Parkinson's.

In essence we need to find a way, a more modern means, of storytelling within our community. We need to create great blocks of knowledge and wisdom that can be recapitulated and built upon. This is how we pass the baton.

BETTER PATIENT ADVOCACY
3: THE DYING OF THE LIGHT

Communication, in any form, is a two-way process. Even a monologue requires a degree of reciprocation. Without this it is meaningless, just words drifting out into space. The monologue requires reciprocity, an acknowledgement that it has been heard even if it has not found favour from a receptive audience. It is still communication and it is still two-way.

Parkinson's is a multifactorial neuropsychopathological condition. No, it's not simply a movement disorder – let's put that one to bed once and for all. Parkinson's is an extraordinarily complex condition or series of interlinked conditions depending on whether one takes a holistic or reductionist position. But the reality of the condition, whichever philosophical stance one adopts, is an arduous day-to-day grind. We can philosophise all we like but the truth is simple. Parkinson's is a nightmare condition, sapping our strength and gnawing at the very sinews of our resilience.

We all know how the story ends. And for many of us, the end of the story is so distant that we may put it aside or prefer not to think of it. We all, to some extent perhaps, mortgage our futures for better todays. The future is unpredictable, the present is at least partly under our control.

But in the same way that the condition has a beginning and an end so too does advocacy. There are few advocates before diagnosis (obviously) and, at the end, equally few. In between, the nature of an individual's advocacy shifts and changes like the sands. In part this is a dynamic woven out of the individual's internal disposition, personal circumstances and the wider community.

There have been several attempts to map advocacy roles as a function of time with some models claiming distinct stages in advocacy. Personally I would stop short of that but I do acknowledge that they can, to some degree, be helpful in an academic sense. But let me tackle the issue of advocacy at a slightly more visceral level, viewing how it projects onto the post diagnostic lifespan of a person with Parkinson's. These are generalities and generalisations so please bear that in mind before putting pen to paper on whether I really meant three years and not thirty three months. Chances are I didn't.

Diagnosis, the starting point of the journey (and know that I really don't like the word "journey" with its connotations of destination) sets the clock running. And often it sends the patient running. To the Internet. To Google and Wikipedia. Fear is swiftly replaced by terror, anxiety by panic, the doldrums by despair. In an afternoon searching on the Internet, one's future is mapped out. And for the most part it's an entirely inaccurate picture, conjured of our darkest imaginings.

Some never get beyond this point, abdicating any future pleasure in an orgy of self pity. And it is easy to do. Indeed most of us have probably been there at some point. And if the Internet were the only source of information, that would be the end of it. Fortunately there are other sources of knowledge, much more positive and creative to help guide the newbies. I am talking of course about patient advocates, role models for the community.

And so it begins. Gradually panic, despair and blind terror are replaced by the purifying sunlight of experience, both personal and collective. Patients terrified by the diagnosis at time zero, learn that they can control some aspects of the illness by accrued knowledge. As time goes on, their own knowledge becomes something bigger, something to be shared. This is the transition from passive recipient of information to informer, from freshman to sophomore.

And before long, the informers become critical informers, no longer simply imparting received knowledge but questioning its authority and forming their own wider view of the condition and its manageability. These critical informers often become opinion leaders, taking their views onto a wider national or even global platform. Often they look back on their former bewildered lives in the year or two post diagnosis with a sense of distance.

Opinion leaders set the tone for communities. Their perspectives on the condition and its context can have huge influence for good or bad.

As time progresses, apathy, the most pernicious of all parkinsonian symptoms, takes its toll. Gradually the informers and opinion leaders fall away. You have to remember of course that the position and status of the opinion leaders is often a reflection of the timespan of their own illness. By its very nature, the condition will have progressed much further in these individuals than in the newbies. The giants may still be there but their voices are quieter. Where once they roared like lions, now they whisper.

As I said before, herein lies the conundrum. The point at which advocates have most to impart is the point at which their capacity to do so is most compromised by the simple day-to-day struggle for survival. And that's not an exaggeration. No matter how brilliant, persuasive and important, it's hard to do anything much use when it takes an hour to get dressed, to eat breakfast and to answer emails. Then it's lunchtime followed by physiotherapy, exercise or whatever. There simply aren't enough hours in the day.

I should declare my perspective at this point. I have had the condition thirteen years. I don't regard myself as a newbie any more. Nor do I regard myself as one of the giants with broad shoulders. I'm somewhere in the middle. I look up to the giants with the same admiration as ever. But now I have to cup my hands to my ears to hear what they're saying. And I watch the brightest and best of the young sophomores building their own communities.

It was ever thus.

The most enthusiastic, the most energised and driven are the ones building the future for Parkinson's advocacy. And they are building it in their own image, a young image. This is all well and good but neglects the vision and knowledge of the giants. And although the numbers are changing, the needs of the older Parkinson's patients are just as important as those of the young onset Parkinson's patients (YOPD). As someone with YOPD (I was diagnosed at forty nine) but now older (nearly sixty two since you ask), I am acutely aware of the separation of the two schools. I'm not sure whether I have 1 foot in each camp or no feet in any camp.

As I said earlier, it's all about passing the baton. And I believe that oral testimonies may well be the route by which the baton is passed. But if it was that simple, I wouldn't be writing. But there are more than one baton. And it's not always clear who is holding. But of course the real question is which baton and who is holding it.

Many readers the first two pieces in this series have felt that I am disparaging of their efforts and blame them for the number of wheels being reinvented. Actually, I don't. I don't believe that they are singly to blame. I feel that the giants are every bit as culpable. Although their voices may be fading, they still have a wealth of knowledge to impart. And like Rumsfeld's unknown unknowns, only they can know the full depth of knowledge.

It is time for them to rage against the dying of the light.

BETTER PATIENT ADVOCACY
4: A NEW MODEL ARMY

Whenever you mention the word advocacy in the context of an illness such as Parkinson's, most people nod in recognition of same. We all know what advocacy is don't we. It's about raising public awareness, right? Well, yes and no. Yes, in the broadest sense it probably is a question of raising awareness but also no, because it goes much further.

But let's at least make this easy and start with the concept of raising public awareness. But we want to do this scientifically, right? So first we need to be able to define public awareness. What is "public awareness" then? And what are its units? We need to know the units. After all we don't measure the speed of cars in kilograms or the size of the Earth's gravitational field in calories. We need to know the units.

Still think it's easy? Well obviously not and at this point you probably feel that I am dabbling in needless pedantry. It may look like that but I'm making a simple point – if you can't measure something then you can't measure a change in that something. And if you can't measure a change in something then you cannot demonstrate that your outcome has been achieved if you can't measure public awareness of Parkinson's then how are you going to show that it has been increased? And, believe me, if you think this is pedantic, I've barely broken a sweat.

I'm constantly impressed by the number of people who have "raised public awareness of Parkinson's" or who intend to do so without the slightest idea of what that might look like. If my objective was to raise public awareness of Parkinson's I would want to be sure not only that the objective been attained but also that this could be described numerically. Numbers are the currency of science. If something cannot be described in numbers, then I would have a hard time calling it science.

This is the very simplest situation and, in advocacy terms, perhaps the low hanging fruit. If advocacy can achieve nothing else, one would at least hope that it could raise public awareness.

Even allowing for liberties in terms of what it is, how it may be measured and defined, there still the great unanswered question of why. Raised public awareness is surely not an end point in its own right but a staging point or a surrogate measure. It is tacitly assumed that raised public awareness will somehow achieve some greater good – raise money for research, influence governmental policy, improve living conditions, maybe even hastening a cure for Parkinson's. This is almost taken for granted. No double-blind studies have been conducted to demonstrate the link between raised public awareness and modified governmental policy. So, if there is no benefit in terms of the things one really wants to change, is raised public awareness a legitimate advocacy objective? Shouldn't advocates be doing something more useful with their time?

I would argue "yes" and, moreover, the best advocates are eschewing nebulous objectives in favour of more direct action. This is to be applauded.

I would argue that if you want more research, then raise money not awareness. If you want to modify government policy, tackle government directly. Raised public awareness counts for little here. If you want a cure sooner, promote better research.

These are more direct means of influencing change and, to my mind, this is what advocacy is or should be about – high-level interactions with high value outcomes. We should be seated on every drug advisory board, every research steering group, every expert panel, every governmental committee. In essence, we should be represented (by ourselves) in every circumstance where decisions are taken that are of direct concern to us.

But I would go further.

Throughout the first half of the 17th century, armies in Britain were geographically constrained, often acting as garrisons or local militia. There was sparse interaction between such forces and their roles were little more than guarding towns. In 1645, Oliver Cromwell formed the New Model Army, a mobile military force of highly trained soldiers constructed around veteran professionals and young conscripts, the old helping to train the young.

This is directly analogous to the current and, I would like to believe, future role of advocates. Currently, advocates are siloed. Individual charities, research bodies and drug companies have their own advocates with their own policies, remits and strategies. In essence militia. Some things are done well, some less well but there is no sharing of best practice. To my mind, we need a New Model Army of advocates and advocacy.

I believe that advocates have much to learn from each other and from their forebears. I believe that, by sharing best policies and actions, it will be possible for advocacy to evolve into a kind of super advocacy. I see this as a natural progression, organic in many ways, and one which best serves the community. We need to share old knowledge from our wiser heads. We need to share new initiatives from our younger brighter minds. And we need to focus on direct, achievable high-level objectives. This means improving the knowledge base and aspirations of the advocate corps. Raising public awareness simply isn't enough anymore. Our new advocates need to be aiming higher.

BETTER PATIENT ADVOCACY
5: THE RETURN OF THE JEDI.

Let's recap.

In part one (standing on the shoulders of giants), I suggested that, although we had unalienable rights to represent ourselves and our interests, we often fell short of our scientific colleagues by failure to build upon existing knowledge in the way that scientists do routinely. I suggested that we needed to take a leaf from their book. In part two (passing the baton), I speculated that we, as a community undervalued and underutilised learnings of past generations and I made the case for storytelling and the oral tradition. Part three (the dying of the light) presented a major conundrum in the sense that wisdom and knowledge gained through experience was at its strongest in the community least able to express that knowledge. In part four (a new model Army), I tried to clarify what I understood by advocacy and some kind of expanded future role for advocates.

As will been abundantly apparent by now, I don't have all the answers. But, like the old scientist that I am, I am nonetheless aware that wisdom lies not in knowing the answers but in knowing the questions. In the words of Charles Caleb Colton, a 19th-century English cleric and writer, "examinations are formidable even to the best prepared for the greatest fool may always ask more than the wisest man can ever answer".

In some respects I am no closer to the answers or the questions. Writing these pieces has been an exercise and an exorcism. I have felt for a long time that patient advocates have important roles to play but that we frequently fail to match qualifications and skills with desired outcomes and objectives.

We stand at a crossroads. Our scientific and clinical colleagues increasingly recognise the value of our input. Whether that is indicative of a sea change in philosophy or simply enlightened pragmatism in the face of such demands by research boards is moot. Ultimately it doesn't matter. For whatever reason, we have been offered these opportunities and we would do well to grasp them.

But to be of greatest service to our community, we need to bring our "A" game. We need also to expand our definition of advocacy. That's always assuming we can define it in the first place. Advocates are powerful instruments of influence. All of the research charities in Parkinson's have advocates in one form or another. Often their roles are focused by the charities themselves but there is a degree to which their roles have a wider, more philosophical dimension. As advocates grow into their roles, they recognise their place in the universe so to speak and evolve into it.

The fresh faced newbies struggling to come to terms with a crippling lifelong diagnosis are, over the span of time, the veterans and broad shouldered giants upon which our knowledge base and understanding are built. We should treasure them. They come in many forms – fighters to philosophers, thinkers and doers, movers and shakers. Each in their own way fitted into the larger jigsaw. Some saw the value of evolution, others of revolution. There were those who wanted to build while others saw demolition is necessary. Some sought to persuade, others to sweep aside. Some become bright beacons to others, rallying points in the darkness. Others flickered and burnt like wildfire, their brilliance sparkling for a brief few moments. Some focused on their own strengths, building edifices of knowledge and experience. Others flitted from flower to flower like butterflies. Some lights are dimmed, while other stars are in the ascendant. I won't name names. For the most part you know who you are. But I will make one exception.

Of all the advocates who have influenced the Parkinson's world, none has had greater influence than Tom Isaacs. No, I don't propose to deify him. He was, and he would admit this himself, an ordinary man driven to extraordinary actions by his illness. In many respects it gave him a purpose in life that he probably felt was missing. I had the pleasure and privilege of working with him for several years. He was inspiring and exasperating. He drove himself hard, aware that he was running out of time (like all of us). He made light of his lack of qualifications when surrounded by highflying academics. But his greatest strength was his ability to get people talking to each other. Oh I nearly forgot to mention – he founded a charity to cure Parkinson's. He always believed in direct action.

Above all, people listened to Tom. They wanted to hear what he had to say. And I think that therein lies the challenge to all of our senior Parkinson's advocates. We need to somehow capture what made Tom so compelling. The senior advocates whose voices are quietly fading away are a resource too easily lost for good. They are like the old Jedi, knights of an older order. They knew about non-motor symptoms years ago. But still, each generation rediscovers them for themselves, reinvents that wheel. We talk about young onset Parkinson's disease as though it was something new. But there were YOPD groups twenty years ago. They came, they went. In the UK, there was Tina Walker, an inspirational leader. She passed away a couple of years ago. And if it were not for voices still alive passing that information on, we would be starting from scratch with yet another wheel.

It's time for these voices to be heard again. They should never have been forgotten. It's time if you will for the return of the Jedi.

I want to end with a list. It's a list of those people with Parkinson's who have influenced me over the years. Sometimes it can be in small ways, maybe a single thing I remembered. Sometimes these are huge influences in my life. I make no distinction in the list below. There is an inherent danger with lists. A danger that one will upset or antagonise those not on the list. It's a bit like a wedding. There will always be some relative who fails to make the cut and trumpets their displeasure widely. Nevertheless, I'm going to take that chance and list those patient advocates who I feel have influenced me over the years. I place here the usual caveat that this list is not complete. Nor is it in any order. Some of show me their inspiration on a single issue. Others have run like a leitmotif through my life. All have contributed something to shape my philosophy of what an advocate is, could be and should be. If you're not on the list, it's probably my fault not yours. Or you may not have Parkinson's.

Tom Isaacs, Jean Burns, Anders Leines, Jill Carson, Shel Bell, Gaynor Edwards, Colleen Henderson Haywood, Eros Bresolin, Simon Laverick, Peggy Willocks, Pete Langman, Linda Ashford, Bob Kuhn, Matt Eagles, Vicky Dillon, Claire Lindley, Martin Taylor, Dilys Parker, Brian Toronyi, Omotola Thomas, Jordan Webb, Andy McDowell, Steve DeWitte, Claire Jones, Emma Lawton, Georg Sternberg, Samuel Ng, Soania Mathur, Larry Gifford, Leslie Davidson, Richard Windle, David Jones, John Humphreys, Reidar Saunes, John Rooney, Les Roberts, Stefan Strahle, Ben Stecher, Karen Raphael, Philip Beckett, David Lohr, Jo Collinge, Ryan Tripp, David Sangster, Heather Kennedy, Tom Gisby, Tim Brandt, Laurie Mischley, Sara Riggare, Elizabeth Ildal, Sara Lew, Mariette Robijn, Rune Vethe, Brian Lowe, Dale Sherriff, Madonna Brady, Jenny Nelson, Phil Bungay, Maria de Leon, Nan Abraham, Niki Oldroyd, Michael Peachey, Alison Anderson, Ian Meadon, Tim Hague, Catherine Oas, Rachel Gibson, Mark Whitworth, Alison Smith, Karen Rose, Kelly Sweeney, Hedley Cox, Bryn Williams, Tina Walker, Nan Little, Sheila Roy, Mags Mullarney , Israel Robledo, Alan Lewin, Margaret Owen, John Silk, Karen Green, Ron Rutkowski, Connie Elliott, Perry Cohen, Kirk Hall, Fulvio Capitanio, Bruce Jockelson, Rachel Clarke, Ray Wegrzyn, Tim Bracher, Kevin Krejci, Briony Cooke and Steve Shea.

SNAKE OIL OR THE ELIXIR OF LIFE –
TELLING THE DIFFERENCE

Let me just say right away that I have a limited tolerance for outlandish claims made without adequate scientific support. Very limited tolerance. In fact I would go so far as to say that I'm practically allergic to pronunciations like "Your Parkinson's cured or your money back! Read Dr Plonker's new book, yours for only $29.99

plus $8.99 postage and packing. Learn how you can beat Parkinson' without drugs" or "The dietary secret THEY don't want you to know" for three payments of $18.99 with a free certificate and advice on hair loss. "Ancient medicine....Secret recipe... Amazon tribes... Natural healing remedy... Cures cancer and piles".

Sometimes the principal selling point is made by hinting at mysticism. The ingredients of this wonder treatment are " mysterious" (nobody knows what the tube contains), "ancient" (nobody knows how long it's been lying there) and the result of the shared wisdom of a lost Amazon tribe (all very well but their cumulative wisdom has still not come up with the internal combustion engine). Don't hold out too much hope for the cumulative wisdom of a tribe that cannot even write their names in the sand with a stick.

Sometimes a product is "natural". Well so is deadly nightshade, sweetheart. And hemlock. "Natural" doesn't mean good. Doesn't mean bad either. We shouldn't forget that many of our modern medicines are derived in some way from natural products as diverse as tree bark (aspirin) and the saliva of the Gila monster (exenatide). Natural is fine in context. But simply stamping the word natural on something is no guarantee of efficacy.

And I get tired of being told at the end advertisements "this product is not available in the shops". Yes, and there are good reasons for that, encapsulated in many acts of retail legislation around the world. Mainly to do with products having to actually work before you can sell them in shops. That's why you have to buy this stuff from some PO Box in the outskirts of Tallahassee and not your local pharmacist. Most pharmacists have a bit of a hang-up with this kind of thing. Hang-up as in "we do not sell this stuff because it doesn't work". They can be sticklers about this sort of thing.

But I must be softening in my old age. I heard myself saying to a friend "well, who knows. There could be something in it" when discussing alternative therapies for Parkinson's. And somehow the act of saying those words made me think not so much that there could be something in it but that, whether there was or not, it was something I should find out for myself. Not reading a book.

Many years back I used to be a Royal Society University Research Fellow (tiny little brag there), before Parkinson's sliced and diced its way through my brain. I mention this because the Latin motto of the Royal Society "Nullius in Verba" translates as "take nobody's word for it".

So with that in mind, I intend to do exactly that – formulate my own opinions on the various alternative therapies available. And here's where you can help. Send me your suggestions. Where should I start? Crystal therapy? Flotation tanks? It's your call. Let me know and I will draw up a plan. And, over the next several months I plan to try out various "alternative" treatments that have been claimed to have benefit in Parkinson's.

Of course I should declare my colours immediately. I am a scientist, have always been a scientist and, I would like to believe, will at least remain a scientist in thought until I can no longer think. So I will be assessing alternative therapies on that basis. Now depending on your perspective, that may be seen as an advantage or disadvantage. If the evidence in favour of one particular therapy is, let's say, a little thin, you can reasonably expect me to seek out those gaps. If I am unconvinced, you will be the first to know. But the converse also applies – if I am satisfied that there is a strong scientific basis for the claims, I will say so.

So if there is a therapy you think stands up to scrutiny, let me know. If you're right, I will sing its praises from the rooftops. If you're wrong, I will attack it like a ravening tyrannosaur.

Can't say fairer than that.

CANADA DIARY 1: THE LONGEST DAY

Up in plenty of time to get to Gatwick. Probably enough time to walk to Gatwick, having been woken, as so often, around four by animals outside. This time it's the sound of copulating hedgehogs in the road outside. Time was when I would have been intrigued by the many nuances of Sonic's cautious courtship, but today, when sleep will be at a premium, I can do without it. After enduring a quarter of an hour of prickly rumpypumpy I invite them to get a room! Those may not have been my exact choice of words.

My driver arrives at seven. We have known each other nearly a decade. We exchange brief pleasantries then hit the road. East Grinstead delays us by forty standstill minutes. Suddenly what was an easy drive becomes a frantic rush against the clock. We roll into Gatwick with minutes to spare. I dash upstairs to the disability desk and pick up a sunflower lanyard to the bewilderment of the staff. Then it's downstairs and a quick saunter through security courtesy of the sunflowers.

I grab a couple of chocolate bars, then make for Gate 51 where I am I am ushered onto the plane. The flight is uneventful. Because of the date – the thirteenth – the plane is mostly empty. I spend much of the flight trying out different seat arrangements – a full length king size bed, spread across the central triple rank of seats., a sort of double length recliner, and, my personal favourite – a kind of modern chaise longue that took me a good half hour and entertained the stewardesses.

The flight to Victoria is much more interesting. A short hop (only fifteen minutes) over the water on a tiny propeller plane, barely climbing above the treetops. The stewardess has only just finished her "Please pay attention to this special safety information" before telling us to "fold your seat table, switch off all electronic devices and prepare for landing". Not even time to take the odd photograph despite spectacular scenery.

We fly over Iceland. I take a picture. We fly over Greenland. I take a picture. We fly over thousands of lakes in northern Canada. I take a picture.

Jill ambushes me in the baggage reclaim. Neither of us can really believe that I'm actually there – in British Columbia at the western edge of Canada.

By the time we get to the house, I'm beginning to get my second wind. We decide, on a whim to catch a little live music in the evening.

My stomach does not know whether it's breakfast or dinner, Friday or Saturday. Fortunately Jill has cooked a huge Mediterranean pie, the size of a manhole cover. And very delicious it is too.

The live music that evening is in the local coffee shop with local (didn't catch their names) musicians, a guitarist and a young dreadlocked girl who might have been her daughter played drums. Think of the coffee shop in "Friends" and you've more or less got the vibe. It's chilled. Welcoming. And good. Just what I need.

When the musicians are finished, they thank people who have travelled far. Jill leaps up, shouts "London, England" and gestures to my half slumbering figure. I have just finished the popcorn and I am not perhaps at my absolute best. But then it is 6am UK time. I think I've done well.

I sleep like a sloth. And dream for some reason of aeroplane seats. And hedgehogs.

CANADA DIARY 2: CASSIE AND MISSY

I know I keep banging on about it but sleep is everything for me. Too little and my waking world disintegrates in a cornucopia of cognitive chaos. Too much and...well, I can't remember when that last happened.

Sleep deprivation is my nemesis and I have reported many instances thereof. So, flying a third of the way round the globe and reeling back eight hours of time difference might be expected to kick things into touch. Right?

Wrong. Well, wrong in this instance. OK I was dozing during the girls' set yesterday but that's hardly unusual - After all, I have slept though some of King Crimson's finest (and most energised and exuberant) musical machinations. Two giggling folkies, an acoustic guitar and an African drum stand no chance.

But last night I had six, almost seven, hours of sleep. Premium sleep. Deep sleep with vivid dreams. Ok, I know the return journey has the capacity to be hellish, but I will have a plan in place by then.

In any case I don't want to talk about sleep. I want to talk about dogs. I am a dog person. Cats don't float my boat. I apologise to my cat owning friends. It's nothing personal. But it's hard to bond with something that doesn't care whether you are alive or dead, except at meal times.

Dogs are different. When a dog loves you, it is unconditional. Sometimes they may be confused by you or your actions but they never stop loving you.

As many of you know, I have partial custody of Louis The Magnificent, as fine a standard poodle as you will ever see. But there are many more types of dog. And today I met, properly met, two very fine hounds – Cassie and Missy, as entertaining a canine double act as you will ever see.

Cassie is the lady of the house. Around six years old, she informs Jill and David, her humans, of every coming and going in the house and her opinion thereof. Not particularly noisily. But firmly. She is affectionate and friendly.

Missy is the young upstart. Six months old and part schnauzer, she is an agent of chaos. Not least for Cassie, who tolerates her playfulness and constant ambushes with remarkable forbearance.

Jill and I spent the morning with Cassie and Missie on a long walk through nearby forest. Dark green, with luminous mosses, and home to a family of bald eagles, The forest is well-managed and very popular among dog walkers. We meet dachshunds, Labradors, schnauzers, Visslers and two huge great Danes.

For an hour or so Cassie and Missy chase each other through the woods. Over hill and dale. It is exhausting just watching them.

We chat with their humans. Those are the rules as any dog owner will tell you. Name. Breed. Age. Potted life story. The usual stuff. We walked for miles. Best part? I totally forgot to use my walking stick. Dogs have that power.

I remember reading once that Catholics believe dogs have no soul and therefore will not go to heaven. But as Robert Louis Stevenson said, "Of course there will be dogs in heaven. How could it be heaven otherwise".

Good point. Didn't say anything about cats though. Sorry

CANADA DIARY 3: TOURIST

A day of tourism covers today's agenda. We park downtown, then head to Canoe, a well respected local Victoria harbour seafood joint, by water taxi, bobbing along against the tide. I choose seared Haida Gwai halibut on petit pois and rösti washed down with a Fat Tug from the Driftwood Brewery. Had I known the Fat Tug was seven percent alcohol, I might have thought twice. Lunchtime drinks are around half that. For good reason.

Four percent is leaning toward robust. Respectable. Five percent is a beer that can handle itself. Confident. Assured. Six percent is a muscular, thuggish brute of a drink. But seven percent is off the chart. A suicide note in a glass. I drink only around half the bottle and still feel I have put up a decent fight.

We lurch up the hill to Chinatown, burping beer and halibut as we go. Within two or three streets, bounded at its northern perimeter by the mandatory dynastic gate, is all of China in miniature. Brightly coloured shops disgorge their wares onto the pavement - fine silks, dried octopi and paper lanterns. A sign points out a narrow ginnel, Fan Tan Alley, named after Fan-Tan, a popular game. The street, less than a meter wide in places, is home to boutiques and jewellers now, a significant distance from the gambling tables, opium dens and brothels that once occupied the alley.

Two other quick stops. Ad hoc souveneering* to buy maple syrup and books on interpretation of pre-Bayreuth Wagner in the psychopolitical landscape of Bavaria. Only a partial success - the shops have no syrup.

Then a brisk schlep across the island to catch the 5pm ferry to the mainland, an hour and a half away. And a change of scenery. From Brentwood Bay to Langley. I emerge from Arrivals and look around. I have forgotten my glasses but, even so, still recognise Bob, arm raised in acknowledgement...

"Have you eaten?" he asks immediately.

I smile. It's going to be a good couple of days.

---

*Souveneering: akin to orienteering but with an additional requirement to collect tawdry geospecific mementoes - chocolate madonnas, tropical snow globes, miniature neon skeletons. That sort of thing.

CANADA DIARY 4: A LESSON IN HUMILITY

On the whole I don't play board games. Sure, I will break out the Monopoly at Christmas or play Trivial Pursuit with the kids. But that's about it. With one exception – Scrabble. I love Scrabble. It reminds me of many happy afternoons playing the game with my mother in her declining years. Nowadays nobody plays the game. Except my friend Bob.

There are three things you need to know about Bob. Firstly, he is an expert Scrabble player. Secondly, if you are around him for any length of time, you'll find yourself playing Scrabble against him. Thirdly, Bob believes actions are stronger than words.

Staying at Bob's house in Langley for a couple of days, I have learnt or, more accurately, been taught a valuable lesson. No, although you could be forgiven for thinking so, the lesson is not to avoid playing Scrabble with Bob. No, the key lesson is not to trash-talk him beforehand.

Knowing that I would be in Canada, and aware of the aforenamed's predilection for the board game, I pre-emptively challenged him to a match, a mano a mano. And had I left it at that, all would have been well. But I couldn't resist a little trash talking, perhaps rather overstating my opinion that I would win.

I should've taken my cue from Bob's response. Which was silence. Not a word. Even when pressed, he would only say that he was "keeping (his) powder dry".

Now Bob and I have known each other for long enough that I should've read the signals. First there was the Scrabble board itself – not your regular version, where tiles slid from one letter to the next. This was larger, with walls to help align letters. Then there was the letter bag, notepad, pencils, scoreboard and so on. This was not the Scrabble board of someone who played once in a while. This was a quasi-Olympic Scrabble board.

Faced with this Colossus, I realized this would be a tough challenge. And so it proved. Within the space of a brace of moves apiece, I realized I was in trouble, holed below the waterline so to speak. My mind went blank. While I feinted with SAD, MUM and POD, Bob parried with MAXIMUM, CRANKY and TACTFUL. Or similar. I forget.

I'd like to suggest that it was a close fought contest, with the lead changing hands all the way down to the wire. But that would be a complete fiction. Frankly it was a bloodbath. Indeed I even think Bob was toying with me, as a cat sometimes does with a mouse, holding back to make a game of it. By the end of the game, I was reduced to hackneyed monosyllables like CAT, BUN and TOP while Bob showboated with TESTUDINARIOUS, PLENIPOTENTIARY and NEUROPSYCHOPHARMACOLOGICAL on a Triple Word score.

(Yes I know those specific words are not usable in Scrabble. I'm just making a point. Don't write in.)

We played in silence. The silence of friends but at the same time, the silence of people concentrating hard on possibilities. Bob's breathing was regular, his tremor absent, his eyes fixed on the board. Only when the coup de grace was administered, and it became apparent that the winning margin was in three figures, did he allow himself a smile. And the twinkle returned to his eyes.

It would be well within the rules of trash talking for Bob to remind me, at this juncture, of my pre-match assertion of victory. It is a measure of the man that he did not, preferring instead to let his actions speak louder than his words.

It was a lesson in humility.

CANADA DIARY 5: BEARS

You think you've seen it all. You think that you know about animals by seeing them in a zoo, digesting the paragraph or two of information in front of their cages, jostling with the crowds to see the more exciting creatures. You read about the natural habitat, where they live, feed and breed. You think you know it all.

You know nothing. Less than nothing.

Until you have experienced animals in the wild, you might just as well read about them in books. I've seen any number of animals in zoos (let's not go there – that's an issue for another day) and been fascinated by successful breeding programs, appalled at inappropriate habitats and uncomfortable in the presence of some, especially the higher primates.

Nothing, but nothing prepares you for meeting grizzly bears in the wild. No television box, picture book or Youtube film prepares you for the experience of meeting these mighty creatures in one of their natural habitats on the Bute Inlet, just north of Orford Bay, country they have shared with the Homalco first nation people for millennia.

The grizzlies collect on the Bute inlet, in August and September, when the river is bubbling with salmon making their way upriver to spawn. As determined as they are to spawn and prolong the species, so too are the grizzlies, determined to have one last fish supper before hibernation.

And yes, grizzlies do hibernate. I believe, without sounding like David Attenborough, that they are the largest animal to hibernate, a process that even today we have made little progress in understanding. The diet of the grizzly is predominantly fruit, berries and roots for most of the year. But in order to hibernate, a grizzly has to increase its body weight by around a third. You can't do that on berries – you need protein. And lo and behold, what should come swimming up the rivers at that time but salmon. Lots of salmon. And big ones at that.

The bears descend on the rivers like crowds at the first day the Harrods sale. Stuff their faces with fresh salmon. Couldn't care less about the berries when there is salmon on the menu.

This single-mindedness ironically helps keep human viewers safe. As long as the bears can smell fish, they will always go for that first. Of course anyone foolish enough to climb out of the bus with a tuna mayo sarnie in hand may find themselves equally attractive. But then, anyone foolish enough to do that should probably be eliminated from the gene pool anyway. And nature can help there.

But assuming you don't try to walk among their number smelling of salmon or wearing a salmon costume (why would you?) You are in actual fact fairly safe. The guides make a point of emphasising the need to stay together and not to use flash photography or any sudden movements likely to spook the bears.

You don't want to spook the bears. They have claws three inches long, sharp teeth and a bite strength that would crush steel. They are also notoriously short-sighted I understand. And they don't like surprises. So as long as you bear these facts in mind, you are (relatively) safe. But at the end of the day, you have to remember that these are wild animals. And big wild animals at that. And you are there with their permission.

The Homalco people run small excursions up the Bute River Inlet. We were a party of seven, bristling with cameras and expectation. Previous excursions had seen very few bears so we were prepared for disappointment. But around the first corner beside a popular viewing point, we saw our first grizzly – a mum with two cubs. On the one hand, a magical sight, on the other a potentially very volatile scenario. Grizzly mothers are notoriously protective of their young, not least from other male grizzlies. If you are admiring the cubs, you do well to keep an eye on the movements of the mother. You do not want to come between the mother and her cubs. Situations can deteriorate very quickly under such circumstances.

We whispered amongst ourselves while the cubs frolicked in front of the mother. It was one of those moments of connection, absolute connection with another living thing. I can't even begin to put into words the sense of oneness with these amazing beasts. There is something absolutely primal about meeting such creatures in their own habitat. I felt blessed that they would allow me in. I will never forget that first encounter as long as I live.

We moved further up country to another favourite viewing location, a bend in the river where the salmon leapt. And the grizzlies were there to offer them a warm grizzly welcome, principally consisting of biting their heads off. In the spawning season an adult grizzly will typically eat around fifty salmon per day.

But perhaps the most edgy encounter was our last that day. Coming round a corner, we noticed a large grizzly male in the middle of the track. We waited a minute or two for him to leave before getting out of the bus. Only then did we notice another bear walking past us below the near riverbank, upon which we were standing, and separated by some scrub brush. He hadn't noticed us and seemed oblivious. But it was closer than we might have liked. The guides had their cans of bear repellent out ready if necessary so clearly they perceived danger. At its closest, the bear was no more than five metres away from me. I stood stock still until he had passed. Only then did I exhale. It was beyond exciting.

After three hours at perhaps half a dozen locations, we had seen thirteen different bears, some young, some old, some plump and ready for winter, others emaciated. The whole of life was here.

We sat in the boat having our lunch. Hardly anybody spoke. We were simply in awe, taken aback by what we had seen. Nature had put on a show. And there is no show more majestic than these magnificent creatures.

As I write, I'm transported back to the riverbank and the sight of the bear cubs playing with their mother. I struggle to articulate the connection I felt with nature and its other inhabitants at that moment. Words fail me. And if you know me, they don't usually.

Outside of experiences with my own cubs, it was the best ever. Simple as that.

CANADA DIARY 6: THE SHORTEST NIGHT

Sunday dawns. After the previous day spent in the North watching bears, today's inevitably anticlimactic. I pack my cases, check and double-check. Everything is hunky-dory – gifts for the kids and from my friends and hosts. I have fallen in love with Canada as much as the Canadians themselves. You would struggle to find a more welcoming people. And now it's time to say goodbye.

I hate goodbyes anyway but goodbyes with good friends are somehow even tougher. I make no bones about it – I'm a soppy old soul prone to blubbing at the least provocation, whether happy or sad. I am booked on the 3:30 flight out of Victoria to connect with the 19:00 from Vancouver to Gatwick. Jill and David come to the airport with me. We decide to have lunch. A well-meaning gesture but one that simply prolongs the agony of parting. The tannoy announces a sixty minute delay for the flight. No big deal. Delays are common place in air travel and the flight to Vancouver is only fifteen minutes anyway.

We hug and hold each other. I teeter on the edge of tears. This has been a magical trip. I turn to the departure gate and walk. I can't look back.

Sixty minutes becomes ninety. Slight concern at this point, no more. Ninety minutes soon becomes two hours. The plane is somewhere near Nanaimo. I do some mental calculations – domestic arrivals and international departures are at polar opposite ends of the airport in Vancouver. Even with wheelchair assistance it will take a good twenty minutes to get from one to the other. In the time it has taken me to make this calculation, a further delay is announced and departure is anticipated at 18:25. With the best will in the world, that will allow me less than ten minutes to get to international departures. The numbers don't stack up. I'm in trouble. I explain as much to the girl on the departure gate. She is way ahead of me, already making calculations and looking at ways of getting me home. Then the plane is cancelled.

This is the cue for meltdown. Or it would be if events were out of my control. But they are not. I have done all my usual preparation and swiftly move to Plan B – return to Brentwood Bay and an offer to cook supper. While I activate Plan B, West Jet start to sort out my alternative travel arrangements. They can get me to Calgary if I want but probably not beyond. I elect to take the same flight out tomorrow with a longer connection in Vancouver.

The girl on the desk is on the receiving end of everybody's complaints. Somehow apparently she's meant to be able to control the weather, lifting the fog in Nanaimo. She is a model of civil professionalism under pressure as people shout and swear at her. In the face of all this needless abuse, I make a point of expressing my thanks for her efforts in trying to reroute me. She bursts into tears.

Take Two.

We are back at the airport the following day. More goodbyes. A strange feeling to be honest. I pass swiftly through security just as the Tannoy announces a one-hour delay in my flight. I laugh out loud, attracting some strange glances from people around me. It must be Groundhog Day.

Turns out it isn't. In any case I have allowed nearly seven hours layover in Vancouver to meet the connection. By the end of that time, I know the duty-free shop like the back of my hand. We board on time. We leave on time. We arrive at Gatwick a good half hour ahead of time.

One of the things that makes eastbound travel so punishing in terms of jetlag is the loss of sleep hours. Going west, you arrive pretty much when you left. Going east, you fit a lot of hours into not a lot of time. It's no wonder they pull down the window shutters. Otherwise the night lasts barely a couple of hours.

An hour later I am at home, unpacking my suitcases. The journey out had been the longest day. The return was surely the shortest night.

THE WINE MERCHANT

For the last goodness knows how many years I have made an annual 'pilgrimage' to 3 rue Sainte Claire in the mediaeval town of Dinan in North Brittany. At that address, behind a modestly unassuming shop front is an Aladdin's cave of wine. For this is the legendary La Cave Des Jacobins, perhaps the best wine shop in France – certainly in Brittany.

Let's face it – buying wine can be a daunting experience at the best of times. It is easy to find yourself out of your depth and not waving but drowning. How often does that happen in your local off-licence in the UK? How many times have you left the off-licence with a bottle or two that cost more than you expected and about which you know less than you anticipated. Against that backdrop of foreboding, wine can do nothing but disappoint.

Not here. Not at La Cave Des Jacobins. The staff are wonderfully helpful, informative and encouraging. If you ask for advice, you get unbiased, descriptive information. If you are neophyte, they encourage rather than intimidate. They want to build relationships not simply have customers. And it is the measure of their success that almost everyone entering the store receives the favoured Gallic reception – lots of smiles and hugs and kisses on both cheeks. Long lost friends reunited. When did you last get that treatment in Oddbins?

Each year, I visit La Cave along with my good friend Peter. Peter is well known to them, a regular customer. Céline and Clothilde greet him like family.

When we arrive, the shop is empty. Vincent has just nipped out to the Bar-Tabac a couple of streets away to buy a lottery ticket – it is a major rollover week in the Euro millions. Céline phones the bar and says that "monsieur Peter est arrivé" before departing to Le KeepFit. Clothilde returns to her accounts. The paperwork for September needs to be finished.

Julien emerges from the cellar, blinking in the light. He sees Peter. More greetings. Peter is in the process of introducing me again when Julien cuts him short, saying that of course he knows me from previous years. It is heart-warming. I only visit once a year but they still remember me. In fairness I do tend to buy a lot of wine – last year Peter and I spent €1100 between us.

The doorbell tinkles and in marches Vincent, brandishing, in his estimation, the winning ticket for the lottery. Clothilde raises both eyebrows, unconvinced. More greetings. More exchanges of family news and gossip. It is a good fifteen minutes before we even begin to discuss wines. Clothilde returns to her accounts.

Peter, along with my eldest daughter and her husband, works with Julien to fulfil his oenological needs while Vincent turns his full attention to me. I present him with my list. He quickly skims through what I have written (I'm amazed he can read my handwriting frankly) and sets to work. Occasionally, he tells me that a certain wine is out of stock – will I accept a better wine from the same region, at the same price. Or, on one occasion, they substituted a magnum where they had run out of bottles. They still charged me only the bottle price.

Eventually, my selections are all completed, discussed and, where appropriate, modified. The wines are boxed up, ready to be taken to the car. Julien asks me about a couple of French regional wines that I have not listed – organic wines from Nizas. I ask how much. "Rien. Gratuit" says Julien.

It's the same every time I go. If they feel there are gaps in my choices, they often fill them with additional bottles, nearly always free. As I said, it's a family operation and they want you to feel part of that. They want to build relationships. And they have been – since 1945.

I gesture to Vincent that I want to take a picture of the two of us but I do not know the French word. "Un selfie" he says. "Ah" I reply "c'est la meme en Anglais". Vincent faceplants.

We load up the car – Peter and I have topped the €1200 mark this time. We hastily work on credible explanations for Julie, Peter's wife. She already regards Peter and I as the worst possible influences on each other!

By the time we are back in Epiniac, we have the story straight and are well rehearsed. "Okay" I say to Peter "The Minervois and Languedoc are fine but we're not going to say anything about the case of Chablis, right?"

Peter nods. I think we may get away with this...

LOSING MY 'VOICE'

In general, I don't know very many proper "writers" within the compass of Parkinson's. Among my own circle I would list Leslie Davidson (obviously), Pete Langman (ditto) and Heather Kennedy (obviously if you know how we both work). There are others but these are the three whose work I read most regularly. And these are the three whose craft most significantly impinges upon mine, such as it is. I have talked intermittently but often with all about the concept of a writer's "voice", our own USP if you will and how it changes over time. Or doesn't.

Some years back I switched from typing my work to dictating it. In large part this decision was taken for me by my progressive decline in dexterity, and consequent increase in the number of typos. What previously had been relatively fluent typing, albeit using only the index fingers of each hand, had begun to deteriorate to the point where I spent as much time correcting as writing. At around the same time, it became apparent to me that speech recognition software had finally reached a point where it was a serviceable alternative to typing. So I made the switch. And after a few simple exercises, essentially reading passages from known texts for half an hour or so, the software became attuned to my way of speaking, my pronunciation and the meter of my speech.

I spoke, it wrote.

But did it materially alter the style of my writing (leaving aside the question of whether it can even be called writing if it is simply the recording and transduction of speech)? I would argue that it did not, simply on the grounds that no reader has ever successfully identified the point where I shifted from one form of verbal input to another. Not one single person has told me correctly when that occurred.

This is flattering. Whatever flutter of wings or flicker of light is caught in my writing, is not noticeably impaired by reading. My writing is robust and determined by what goes on in my head rather than my fingers or larynx. In essence my 'writer's voice' is located entirely within my mind and not my hands. This may seem obvious, but for me, with a relatively wordy and free-flowing style, this is very satisfying. I would hate to think that the style was conditioned by the means of its transfer to paper.

This probably doesn't matter one jot to most of you but writers get bothered about this sort of thing.

As I said, nobody spotted when I started using my speech software. Indeed those very few who had an inkling, often say that any perceived or imagined style change is for the better, a less staccato and more cantabile format. The greatest compliment people pay me is to say that they can hear me in the writing, that in essence they imagine me reading to them. I find that on the one hand very generous praise and, on the other hand, an intimation that writing is somehow little more than chat. I know that Pete, Leslie and Heather would disagree.

Let me change tack for a moment. Bear with me, this will make sense.

Every year, around December, I have an appointment with my speech therapist. The pattern is always the same. We chat for a little while, she takes decibel measurements and then tells me that I'm wasting her time. Then we chat some more, about our children as often as not, before making another appointment a year hence.

I take great pleasure each year in being pronounced a waste of time. Because being a waste of time means that my voice has not deteriorated in the intervening twelve months.

This is of more than idle interest. Many people with Parkinson's have vocal problems of one form or another. Some stutter. Some can sing but cannot speak. Some have voices barely audible over the background hubbub. Some slur, some whisper. Hardly any shout, or can shout. Some have difficulty with certain words or sibilants. Some trip over one word but not another. Some cannot separate syllable multiples. If you are a neurologist or a speech therapist, the average Parkie will keep you occupied with a rich vein of symptoms.

I mention all this because recent events lead me to conclude that my voice – my physical voice – is changing and that in turn is altering my voice – my writer's voice. It's nothing major and, to others battling far worse consequences of their Parkinson's, may seem trivial. But three things, in conjunction, lead me to this conclusion.

Firstly, and I've noticed this over the last few months, I am often asked to repeat myself. This suggests that my voice is either weaker or less precise. As I said, nothing major. No catastrophic decrease in volume or clarity, just a few subtle differences which mark the boundary between comprehension and misunderstanding.

Secondly, Alexa keeps playing the wrong music. I ask it to play, say, "Big Log" by Robert Plant and it will offer me a medley of music for the Paraguayan nose flute. Or a request for the second movement of Brahms's third Symphony will be turned down in favour of the International Scout movement playing Cumbaya on recorders and kazoos. I'm all for open-mindedness in music but there is a limit to how much of this I can take.

Thirdly, and perhaps most significantly of all, the very speech recognition software I'm using to dictate this is beginning to mishear (or misinterpret) me. This is particularly disquieting since I've already had to forego typing, except on rare days when the meds seem to work according to plan.

On its own, these changes are dispiriting. For two reasons. Firstly, there is a serious chance that my speech therapist will begin to take an interest rather than simply tell me I'm a waste of time. That itself will mark a significant point of transition. But my biggest concern relates to Deep Brain Stimulation. I'm on a list to be considered for DBS at the moment. There is inevitably a significant waiting list so there is unlikely to be precipitous action taken. But one of my concerns relates to speech. DBS is excellent for treating tremor, perhaps my most debilitating symptom, but it does so at a price. And that price is often impairment of speech to some degree, especially when the electrodes are targeted on the subthalamic nucleus rather than the globus pallidus, as is the current fashion. In some people it is barely noticeable whereas in others the disruption can be quite severe, rendering the normally well understood nigh on incomprehensible.

That would be the supreme irony – that DBS takes away my spoken "voice" but improves my tremor to the extent that I could type again, in essence giving me back my typed "voice".

Five hours of complex neurosurgery to achieve that – oh how I would laugh.

TAKE MY ADVICE

I have had Parkinson's now for thirteen years, some better than others. I've had it long enough that... well, I've had it long enough, let's just say.

But one of the blessings (or curses, depending on your demeanour) of reaching such an unenviable milestone is that I am often asked for advice about living with Parkinson's. And I should say immediately that this has nothing to do with any great scholarship on my part. It's more a reflection of the fact that I'm still here.

As you get further along in Parkinson's, the numbers begin to dwindle. Five years into the mission everyone is imbued with enthusiasm and a "can do" mindset. You march along, at the head of a column so to speak, breathing in the fresh air and confident in your cause. Then after a decade, you look over your shoulder and there don't seem to be as many. Another couple of years and there are fewer still. By the time you get to twenty years post diagnosis, you have reached godlike status.

Well I certainly haven't reached the status of any deity, except perhaps the Buddha, and that merely an observation about my increasing belly. But I do get asked a lot for advice about this miserable condition. Nothing special there. We all get asked advice at some stage. The question is what do we do.

That may give you a couple of scenarios.

First example: Tristan, a friend I have known for many years calls me up and tells me that he is experiencing troublesome dyskinesias and what should he do about it. I have a think and then give him my advice.

Second example: Gunther, somebody I met once at a conference many years ago (couldn't remember his surname) emails me and asks me for my advice on dealing with dystonia for instance. I am busy but manage to send him a couple of paragraphs of thoughts.

How do these two situations differ? In each case, a request for advice has been made and in each case advice has been given. But is the character of the advice the same? Does it have the same strength and importance?

The closer you inspect the two scenarios, the more pronounced are the differences. In the first case, Tristan was asking me for my advice as a friend. I gave my advice openly without thinking too deeply. It was a casual question and perhaps received a casual response. In the second situation, Gunther was at best an acquaintance. My response to him was understandably a little more circumspect and less direct.

So what, I hear you say. All advice is the same. It doesn't matter who asks and who gives advice.

Actually it matters a lot. Let me explain.

The first scenario is easy. A friend asks for advice lightly and takes it in the same vein. If my advice doesn't help or, perish the thought, makes things worse, Tristan will tell me so and I will have another think.

But the second scenario is different not only in the way the advice is requested but also in the way it's received. Perhaps Gunther has heard that I'm a neuroscientist and therefore anticipates my advice being more valid. There is thus a higher expectation of good advice and, in equal measure, a greater disappointment if my suggestions fail to bring relief.

Let's go further.

Let's imagine that things go really badly and Gunther finds himself in hospital after following my advice. He feels aggrieved that the advice of a neuroscientist should be so dangerous. He writes as much on one of the many Parkinson's bulletin boards and chat rooms.

In his diatribe, he draws attention to the poverty of the advice given by myself, feeling that it falls short of that expected of a neurologist. It probably does and therein lies at least partly an explanation. A neurologist and a neuroscientist are not the same thing. I know that and so does every neurologist. But evidently Gunther did not. He took my advice to be that of a fully fledged neurologist rather than a semi retired neuroscientist.

Does it matter whether I am a neurologist or a neuroscientist?

Yes. It matters a great deal. Knowing that I am a humble jobbing neuroscientist, Gunther would perhaps hold on to my advice more lightly, viewing it against a backdrop of similar advice from others. However, labouring under the misapprehension that I'm a neurologist, his expectations are much higher and disappointment more profound.

However, even there, he has little source for complaint. He has received free advice. And, as we all know, things are worth what you pay for them. Had he paid for the consultation with a genuine hard-boiled neurologist and then found himself in hospital, his indignation would be justified.

The point I'm making is that the character, import and meaning of advice is contextual. It depends on who is asking and who is receiving information and the relationship between them.

Let's go back to Gunther for a moment. Still in hospital, he is grumbling about the poor advice he has been given when, discreetly, the man in the next bed passes him his business card. "Mr S Beckmesser, Lawyer".

I think we can all see where this is going.

I have deliberately painted an extreme situation. Or it might seem so. But the truth is that, in an increasingly litigious society, we may find ourselves facing litigation for poor advice, no matter how honestly and earnestly it is given.

The biggest danger as I see it lies in the forums and chat rooms where advice is liberally dispensed amongst total strangers. Somehow it is felt that the cloak of anonymity allows expression of ideas that may be dangerous. It is not even unknown for trolls to deliberately dispense dangerous advice.

There are a few things more calculated to unsettle the administrators of chat rooms than the perception that they may be held responsible for the consequences of any rogue postings or honest but dangerous advice. Not surprisingly, conditions, caveats and codicils abound.

It is increasingly difficult to know what is reasonable. On the one hand one does not wish entirely to suppress novelty, creativity and endeavour on the altar of litigation. On the other hand, perhaps there is a need for people to emphasise their credentials (or lack thereof) and context before giving advice.

Personally, I am more reticent about giving advice these days for some of the reasons outlined above. A pity really because, after thirteen years, I have plenty of advice to give.

JAMES DEAN R.I.P. (PLEASE)

Perhaps the most bizarre news I have read recently is the announcement by Magic City Films of their plan to film "Finding Jack" based on the novel by Gareth Crocker. On its own this would be something of a non-news item. Another Vietnam film. Yawn. Isn't it time to put that subject to bed?

But Magic City Films have found a way to make that seem newsworthy. The stroke of genius? To cast James Dean in a supporting role. Yes, that James Dean.

James Dean was born in 1931, making him currently 88. Of course, extreme age is no bar to acting. Look at Vincent Price, Christopher Lee and so on. No, age is no impediment to acting. The principal objection to James Dean taking the role of Rogan in the forthcoming film is simple.

He's dead.

And he's been dead for a long time.

To be more specific, he died in 1955 in a motor accident.

As far as I know, no attempt has been made at any stage to clone him. Nor, unlike Elvis, is there any significant support for the notion that he faked his death. I have yet to see my first "James Dean is alive" T-shirt.

No, as far as James Dean is concerned, dead means dead.

So how exactly are the filmmakers going to get round this? I'm not sure how many modern actors would be comfortable acting opposite a dead person. Having said that, there are probably plenty of modern actors who are as good as dead, such is the premium that the movie industry places on looks over, say, any acting ability. Who knows – maybe a dead James Dean is probably a better actor than a living... (Fill in the name of your choice).

Having said that of course I should acknowledge that star quality is only a vague reflection of acting ability. Take

Humphrey Bogart for instance – one of the more wooden actors of the silver screen, and a shortass to boot. Yet he was also one of the most charismatic stars of his generation. Casablanca remains my all time favourite film.

That still doesn't explain Dean's casting, sixty four years after his death.

The truth is that this is not the actual Jimmy Dean. Well, not in any form that we would understand. This is a mess of electrons converted into some kind of visual representation of the great man. Apparently he has been digitised. His every facial gesture has been rendered to a series of coordinates. Every sneer, snarl, smile and squint reduced to numbers.

I saw the results of this kind of process in 2016 in Rogue One, a Star Wars spin-off. Peter Cushing was berating some other minor character over inappropriate use of Death Stars or running in the corridor. Or he would have done, had he not died in 1994. No, this was not old footage of Peter Cushing. It was his electronic stunt double. Or something like that. And to be honest he was pretty good, if anything rather more animated in death than he had ever been in life. But then of course his long and glittering career with Hammer films lent him a certain authority when portraying the dead or undead.

Apparently Dean's family are okay about having him electronically resurrected. Well, they haven't actually said that they don't like it. So maybe that's the same thing. Personally, I don't know whether to express my condolences or congratulations. I'm sure when his family buried him so long ago they cannot have envisaged his electronic exhumation. Can we really be sure that Dean would have taken this role?

The irony is that James Dean is immortal on celluloid. Bringing him artificially back to life on the cinema screen just might kill his legend off for good.

LEARNING TO COUNT MY BLESSINGS

The approaching Christmas season or, as my younger daughter persists in calling it "The Most Wonderful Time of The Year" invariably hooks me back to my childhood. My mother, having lived through a world war, felt that Christmases should not be taken for granted. She never tired of telling us (my sister, brother and myself) that rationing was severe during the Second World War and that we were able to enjoy Christmas now solely because Hitler had been defeated. To this day, I have been unable to fathom the exact relationship between the successes of the Africa Corps and portion sizes for plum pudding at Christmas.

There were certain grown-up phrases which came repeatedly to mind. One of her favourites was "Christmas is about giving not receiving". That cracked me up every time. Sometimes I would even laugh openly when she said it, so absurd was the notion. After all, every Christmas for as long as I can remember I had received gifts aplenty. Train sets, tricycle, battalions of toy soldiers, a diecast model Luger and so on. And each year I prepared a long list of things I wanted for Christmas. I say "wanted" but what I really meant, of course, was "expected". In one sense, my mother was right – Christmas was indeed about giving. Specifically my parents giving. To me.

Now before you say anything, or whisper words like "brat" and "spoilt", let me reassure you that, over the subsequent half-century, I have reached different conclusions about the comparative merits of giving and receiving. And I can say with my hand on my heart that, this Christmas I am not expecting a train set, tricycle, toy soldiers or any a sort of handgun, German or otherwise.

But then there was also my mother's other favourite "the best things come in small packages". As with the spiel about giving and receiving, this too was comically absurd.

How exactly was the box my sister received, little larger than the Sindy doll it contained, meant to match up to my Action Man Armoured Personnel Carrier, which featured engine sounds and a functional turret gun and came in a box the size of a tea chest? So large a box in fact that my mother gave up wrapping it halfway, muttering about being unappreciated. She was. Just like other mothers. Why is it that we only learn to appreciate them once their work on earth is done?

I was seven then. I'm sixty two now. And I acknowledge that these may have seemed greedy, and my expectations those of a spoilt child.

You would be correct of course. I was spoilt rotten. And in fairness, nothing much had happened in my life to dispel this rampant sense of privilege. Little Lord Fauntleroy I wasn't. But I did still expect to be showered with gifts at Christmas. And, if I'm honest, I think my parents gained some vicarious pleasure from giving their children things they never had.

Gratitude is not a simple construct. Young children often have little sense of gratitude – their emotions being essentially "happy" or "sad", the basics. They have no concept of "thankful". That takes time. But there are good reasons – and I mean neuroscientific reasons – why gratitude is meaningful and important.

Gratitude, and the expression of gratitude, have some quite neuronally specific effects. We may feel that gratitude gives us a warm feeling all over but that's not the case in the brain. Functional MRI studies show that the expression of gratitude causes particular activation of the medial prefrontal cortex [Kini et al, 2016]. Individual differences in the expression of gratitude also revealed differences in the fMRI scanner. Liu et al (2018) found some changes in the right middle occipital gyrus extending to posterior superior temporal sulcus and temporoparietal junction.

This probably means precious little to the man in the street – if I'm honest it doesn't mean a whole lot even to me – but the point is that there are specific neuroanatomical differences between people who express little or much gratitude. And there are long-lasting neurochemical changes incurred by the expression of gratitude. And if that were not enough on its own, there is even some evidence that expression of gratitude makes heart rate lower and more stable. Gratitude, and the acquisition of gratitude, therefore probably has long-term benefits on health.

Gratitude is not a simple emotion and requires the activation of several brain areas in concert. We are just at the beginning of discovering how gratitude may have general health benefits but if you can say nothing else, one thing is clear. We should count our blessings.

Kini P, Wong J, McInnis S, Gabana N, Brown JW. The effects of gratitude expression on neural activity. Neuroimage. 2016 Mar;128:1-10.

Liu G, Zeng G, Wang F, Rotshtein P, Peng K, Sui J. Praising others differently: neuroanatomical correlates to individual differences in trait gratitude and elevation. Soc Cogn Affect Neurosci. 2018 Dec 4;13(12):1225-1234.

APPENDIX

THE BAYREUTH MASTER

The Wagner scholar and philosopher Brian Magee once stated that great music was music greater than it could be performed, or words to that effect. In essence his thesis was that the music was greater than could be mirrored in a single performance. There were many different ways of interpreting the music, each equally valid, but none representing more than a partial view of the work.

Nowhere is this more true than in the work of Richard Wagner in general and, specifically in his colossal masterpiece Der Ring des Nibelungen. Even by the standards of Richard Wagner, and he wrote some very large operas, The Ring is monumental. A trilogy of operas with a preparatory *vorabend*, the work is on an unprecedented scale. The *vorabend* itself, a hundred and fifty minutes without an interval, is longer than most operas. Most think of The Ring as a tetralogy.

In the hundred and forty years since its premiere at Bayreuth in 1876, no work has been more analysed, interpreted, misinterpreted, championed, vilified, adored, hated or studied than The Ring. One thing it has never been is ignored. Wagner's music and especially that of this tetralogy stands as a monolith over the 19th-century, influencing philosophy, drama, music and politics ever since.

The music of Wagner has been adopted, or more accurately misappropriated, by politicians to their own end. Especially right-wing politicians. And there is no sadder association in this respect than with the ideology of Adolf Hitler. Wagner was no right-wing politician. He espoused left-wing ideals and revolution. Apart from his lamentable anti-Semitism, his politics were allied more with the left than the right. And even his anti-Semitism came second to art – the world premiere of Parsifal was conducted by Hermann Levi.

It is one of the greatest tragedies of mankind that a deluded Austrian corporal should have built his own perverted ideology on the flaws rather than the strengths of Wagner's character. His mistaken grasp of Wagner's key philosophy destroyed an entire world order. We should not be surprised that our perception of Wagner is tainted by this association.

Normally I would not discuss politics in the context of opera but The Ring is such political opera that it is impossible to ignore. At one level the tetralogy is a gigantic fairytale saga of gods, men, giants, dragons, dwarves and heroism. And the operas survive perfectly well as nothing more than that. It is easy to stage a production of The Ring that concentrates on the magic and good old-fashioned storytelling. For many, that is the way into the world of The Ring, an easy way of dipping one's toes into the visual and sound world of Wagner.

But The Ring is so much more than fairytale. The Ring is a political drama. Alliances are forged, truths are told, friendships are betrayed on the altar of expediency, power is widely used and equally abused. The events of this mighty drama, stripped of their winged helmets and horns, are as relevant today as ever. The drama of Gotterdammerung has been played out in Bosnia, in Somalia, Rwanda and elsewhere. The Ring is as meaningful today as ever.

The Ring is also an exploration of human psychology. How are human beings motivated to take one course of action over another? What drives and urges determine action? How do people balance short-term and long-term benefit? Do human beings truly behave altruistically? Where does personal benefit end and collective good begin? How are choices made and factors weighed? All of this and more forms part of the fabric of The Ring.

Interwoven throughout this complex fabric of fairytale and politics is a very human story of love in its many facets. There is love at its most conventional, represented by marriage, on the one hand. There is the forbidden physical love of brother and sister. And there are all points in between. The Ring is as much an exploration of love as anything else.

To regard this tetralogy as somehow irrelevant to our current times is to betray, at worst, a fundamental lack of understanding of the work or, at best, a disinterest in life and its many features. There is no work of modern art to parallel Der Ring des Nibelungen.

RECORDINGS OF THE RING

Despite the jawdropping scale of the tetralogy, record companies have not been slow in recording the work. The CD catalogues are full of recordings, both live and studio, of the work. Old recordings have been exhumed, tidied up with modern digital technology, and released in sound which comes closer than ever to the original. I have heard recordings made in Bayreuth more than a century ago sounding fresh as a daisy.

So let's suppose for the sake of argument that you are looking to buy a recording of The Ring cycle. Which should you choose? Should you indeed choose one or perhaps rather pick individual recordings for each of the four operas? There have been some magnificent recordings of individual operas that have either been recorded as stand-alone pieces or where the remainder of the tetralogy is either lost or was never recorded. Some of these are remarkable recordings in their own right. I have always greatly enjoyed the recording of Die Walkure with largely London forces under the direction of Erich Leinsdorf from 1962, a recording which threatened to derail the Culshaw/Solti Ring. A wonderfully brisk and terse account of the opera, capturing the galloping pace of the drama, but one which never spawned an entire cycle. Similarly I would not wish to be without Karl Elmendorff's recording of Gotterdammerung from Bayreuth in 1942. The notion of the twilight of the gods, just as the war on the Eastern front began to turn against the Third Reich, can hardly have been lost. The recording has a reflective, almost valedictory tone. And I don't think it's just me and my imagination. Both are glorious performances and even though more than 70 years old, the music shines clearly through.

But these are individual operas. Let us put aside the notion of individual operas for the time being and assess The Ring as a single work. Which recording then should you buy and why?

Rather than review each recording chronologically, I will try to present my thoughts and comparisons in a more engaging way. For me, the discovery of The Ring cycle in 1974 was the discovery of Georg Solti as a conductor. To hear Solti's Ring was to be transported to an entirely different world. And Solti brought drama to The Ring in a way that few conductors before or since have managed. Much is written about the architecture of The Ring and I will come onto that later but nobody captured the huge tidal waves of sound quite like Solti. For him The Ring was very much a magical fable and his exposition of the music is almost in that vein. Siegfried's funeral March from Gotterdammerung is as dramatic and brutal a portrayal as has ever been recorded.

Solti was helped by legendary singers albeit often captured late on in their careers when the voices were on the wane. Nonetheless, to hear some of these great voices in gloriously recorded stereo was wonderful. Jon Culshaw's book "Ring Resounding" describes in detail the recording of The Ring over the course of eight years and the machinations necessary to get so many egos into one room. It is a testament to Culshaw's unique managerial abilities that The Ring recordings have a cast that includes Kirsten Flagstad, Birgit Nilsson, Wolfgang Windgassen, Gottlob

Frick and, perhaps greatest of all, Hans Hotter. To hear Hans Hotter delivering Wotan's farewell to Brunnhilde at the end of Die Walkure is to hear opera at its absolute finest. Sure the voice is wobbly and certainly struggles with some of the intonation. He was recorded some time after he had retired from theatre but the sheer majesty of his performance is breathtaking. Nothing could speak more eloquently of the humanity of The Ring than this.

In many ways, the Solti Ring was a time capsule, the final summary of a glorious age of Wagnerian singing. And for that we should be profoundly grateful.

To be honest, the recording has not aged well sonically. I remember it from my youth as a glorious wall of sound and it still is of course. But the transfer of this old analogue recording to digital has highlighted some of the sonic flaws – the tendency to distort or saturate at very high volume for instance. Ten years ago, I would have picked this recording is my first choice – my desert island discs – without hesitation. Now I'm not so sure and I'll come back to that.

Not long after Solti finished his recording of The Ring in Vienna, Herbert von Karajan started to record his statement on The Ring. Conceptually this could not have been more different. Where Solti drove the drama along on tsunamis of orchestral sound, Karajan reduce the orchestration almost to the level of chamber music when supporting the singers. For him the words were the priority and his sound allows the details of the orchestration to be heard as never before. The recording was made in Berlin and the dryness of the orchestral timbre is further helped by the absence of echo. This more than any, and it pains me to say so, is a style of presentation that would resonate particularly with Wagner himself who constantly railed against impresarios reluctant to modify their theatres to accommodate his works. Some singers reprise their roles from the Solti recordings. Gerhardt Stolze allows himself even more sprachstimme under Karajan than for Solti. But for the most part this is a fresh group of singers, sometimes without explicit Wagnerian pedigree, but recorded for their musical intelligence. At the end of the day, this is Karajan's vision of The Ring and singers are secondary to that.

For many listeners, you have to go back further and closer to home for the authentic Wagnerian sound. By "closer to home" I mean of course Bayreuth, spiritual home of The Ring. But the same phrase also refers to the doyen of post-war Wagnerian conducting, Hans Knappertsbusch. Many of Knappertsbusch's recordings are available in some form or another if you really dig, but perhaps the most easily accessible is from Bayreuth in 1956, a mono recording in adequate sound.

Although the Bayreuth recordings can always claim primacy of authenticity – the master did after all build the opera house specifically for The Ring – we should not forget that the first two Ring operas were premiered in Munich and thus subjected to a different acoustic.

By modern standards Knappertsbusch is a dinosaur and, in many respects, the reason people resist Wagner. His tempos are slow by any mark, and his emphasis is very much on the orchestral architecture. In the longer passages, such as the dawn sequence from Gotterdammerung, this works well. But for modern ears there is also the sense that he does not reflect some of the intensity of the drama. Unlike the volatile Solti, Knappertsbusch is unmoved by stage drama. For him, the drama takes place in the orchestra pit. That said, the 1956 recording has some glorious singing, not least by Hans Hotter, Wolfgang Windgassen and, Astrid Varnay. Hotter and Windgassen's voices are less strained than they were a decade later under Solti and Varnay is electrifying as Brunnhilde. She didn't always hit the notes absolutely accurately but her vocal range was magnificent.

In many ways the singers are the reason for buying the set. The orchestral tone is adequate but softened by the

Bayreuth pit. Fascinating though the orchestral sound is, Knappertsbusch stamped his authority all over it. I used to be a much greater fan of his conducting but nowadays I'm a little less than certain. The world has moved on from this lumbering orchestral leviathan. Still, judge for yourself.

Although Knappertsbusch represented the ponderous splendour and majesty of Wagner's music, others sought to extract subtler textures from the orchestral tapestry. None did this better than Wilhelm Furtwangler. And as if that was not good enough in its own right, we are spoilt for choice. In his career, Furtwangler recorded The Ring not once but twice under different conditions. Both performances are essentially live and in the case of the La Scala Ring, made during public performances in 1950. The second set, for Italian radio in 1953, was also recorded live in the sense of a single performance albeit without an audience.

Live recording suited Furtwangler better than most conductors. Furtwangler had a way of extracting breathtaking performances in front of an audience and orchestras found themselves caught up in this electricity. What we see in these recordings is the most natural, unhurried, but excitable orchestral playing. Furtwangler modified tempo better than anyone, subtly and credibly, in order to point out drama but always without exaggeration. If it was there in the music, Furtwangler would offer it. If it was not, he would not create falsehoods. Tempos are natural and generally faster than Knappertsbusch.

Of course there are always drawbacks to live performances. Mostly these are the audience. The 1950 recordings are particularly bad in terms of audience "participation". By modern standards, the amount of coughing is shocking. Italian audiences are notorious for their disrespect, but the continuous coughing is distracting. In some passages, the singers are having to cope with what sounds like a tuberculosis epidemic. The 1953 set has no audience and although the performance is live, there are no such interventions. For some this will be a deciding factor.

So what distinguishes the two performances? Well, the first thing to say is that the orchestral sound is noticeably better in the 1953 recordings. Both are mono but, whether due to more recent engineering or better raw material, the 1953 recording is clear whereas much of the detail on the 1950 set is lost in the continual coughing. It's really that bad.

For some, these aspects may be sufficiently damning as to preclude consideration of the 1950 set. But that would be to forget the singers and make the assumption that The Ring is only the orchestral sound. When we turn to the singers, the margin between the two sets is much narrower. Because in 1953 Furtwangler's Brunnhilde was none other than Kirsten Flagstad, in her prime. And we should remember that she was, for many, THE Brunnhilde of her generation. A voice of absolute clarity, tonal accuracy and awesome power, she was simply awesome. I never heard her in the flesh but her many recordings are testament to this vocal colossus.

The remainder of the cast is certainly good – Ferdinand Frantz as Wotan and Max Lorenz as Siegfried for instance – but this is really the Furtwangler Flagstad show. Although marred by the audience this is still epic singing and performance.

Moving on to the 1953 Italian radio recordings, the listening is easier. No non-musical interruptions to break the spell. Just luminous conducting by Furtwangler with what would be a very fine "cast performance". By that I mean that the performance is very much a company effort rather than a vehicle for a single outstanding singer. Perhaps I'm deluding myself, but that seems to me to be reflected in the performance. There is no sense of anticipation of the big arias which are simply integrated into the music. Wotan is again Ferdinand Frantz but Siegfried is now Ludwig Suthaus, an excellent singer but no Max Lorenz. Brunnhilde is Martha Modl, a Bayreuth favourite. She is no Flagstad, but then nobody else was. She swoops upon the notes like the Valkyrie she is and although not always accurate her voice is always dramatic and compelling.

Two recordings by Furtwangler, both compelling but in entirely different ways. If only we could have the best of both sets in one. But then who would agree that? My own preference is for the 1953 Rome set. An integrated performance with all Furtwangler's electricity and no distracting audience participation.